W0235080

Politics and Policies in the Debate on Euthanasia

Inês Santos Almeida • Luís F. Mota

Politics and Policies in the Debate on Euthanasia

Morality Issues in Portugal

Inês Santos Almeida
Research Unit on Governance
Competitiveness and Public Policies
(GOVCOPP)
University of Aveiro
Aveiro, Portugal

Luís F. Mota
IJP-IPLeiria - ESTG
Instituto Politécnico de Leiria,
Polytechnic University
Leiria, Portugal

member of GOVCOPP
Aveiro, Portugal

ISBN 978-3-031-44587-3   ISBN 978-3-031-44588-0   (eBook)
https://doi.org/10.1007/978-3-031-44588-0

This Palgrave Macmillan imprint is published by the registered company Springer Nature Switzerland AG.
The registered company address is: Gewerbestrasse 11, 6330 Cham, Switzerland

Paper in this product is recyclable.

ACKNOWLEDGEMENTS

Just like the legalization of euthanasia in Portugal, the process of writing this book was also long and complex. When the challenge to write this book arose, we thought it would be easy and quick. Inês's master dissertation had already covered the analysis of the first round of debates on euthanasia, and we thought it would just be a question of updating the analysis for the second round and for a quick third round that was about to start. Little did we know that the legislative process would last two more years and so would constant need to update our analysis, aggravated by challenging professional and health periods in both our lives.

But these difficulties did not discourage us, and we managed to complete this goal. This would not be possible nevertheless without the support of some people.

First, we would like to thank the editorial team at Palgrave Macmillan for continuing to believe this endeavour would come to an end and for extending the initial deadlines.

Second, we would like to thank all the colleagues who provided comments on our ongoing study of morality policies in Portugal in general, and of euthanasia in particular.

Third, we express our appreciation to the representatives of political parties and civil society movements who agreed to be interviewed and spend some of their time sharing their views on the legislative process with us.

But our greatest gratitude goes to our families, for their continuous support and patience in this and other academic endeavours.

On an institutional level, we would also like to stress this work was financially supported by the Research Unit on Governance, Competitiveness and Public Policy (GOVCOPP-UA) (UIDB/04058/2020) + (UID P/04058/2020), funded by national funds through FCT—Fundação para a Ciência e a Tecnologia—and by a research fellowship attributed by FCT to Inês S. Almeida (2021.06378.BD).

CONTENTS

LIST OF FIGURES

CHAPTER 1

Introduction

Abstract In this chapter, we start by briefly explaining why the topic of euthanasia is a very divisive issue and may be considered a morality issue. Moreover, we present our object of study—the political and public debate on euthanasia in Portugal held between 2016 and 2023, which ultimately led to the legalization of active euthanasia in May 2023 (Law 22/2023) and, consequently, Portugal to become the eighth country in the world to do so.

We also justify the relevance of this research object, which results from the fact that Portugal is often considered to have a religious and conservative society but has had a 'permissiveness wave' regarding morality issues since 2001, which remains understudied. Taking into consideration our goal of analysing who were the most active political and social actors in the debate, and the arguments they used against or in favour of the decriminalization of euthanasia, we explain our methodological strategy and our sources of information. We finished the chapter with an outlook of each chapter.

Keywords Euthanasia; Portugal, Methodology

I. S. Almeida, L. F. Mota, *Politics and Policies in the Debate on Euthanasia*, https://doi.org/10.1007/978-3-031-44588-0_1

1.1   The Research Object and Its Relevance

The possible legalization of euthanasia is a very divisive issue in most, if not all, societies, as it deals with fundamental moral, ethical, philosophical, religious and even legal questions about something as important as life and death (Preidel & Knill, 2015, p. 79), which are part of people's deep core beliefs and often affect their preferences on public policies (Weible & Nohrstedt, 2012). Euthanasia can thus be considered a morality policy, a (sub-)type of public policy which includes several issues such as abortion, same-sex marriage or even prostitution or drug use since discussions about them often mobilize several morality-related arguments, rather than solely material or rationally-oriented arguments (Heichel et al., 2013; Knill, 2013).

Besides being a very controversial issue, euthanasia is also a somewhat complex issue, not only because there are different types of euthanasia which are often confused—for instance, there is a difference between passive euthanasia, assisted suicide and active euthanasia, which can be considered as increasing levels of permissiveness—but also because it implies the assessment of very complex criteria—for instance, to determine what constitutes a condition of suffering of great intensity that may justify death (Preidel & Knill, 2015).

Despite its controversy and complexity, the possible legalization of euthanasia has been discussed in an growing number of countries, particularly since the 1980s, as also happened with other morality issues (Preidel & Knill, 2015). As Engeli et al. (2012a) mention, the increasing dispute on morality issues is mostly related to trends of secularization and individualization in Western societies. Preidel and Knill (2015) also consider this assumption to be valid for euthanasia, adding the relevance of the advances in medicine which enable people with severe illnesses to have their lives prolonged, which was not the case before.

The debate on active euthanasia held in Portugal in recent years is an example of this trend of discussions, even though it took place a few decades later than in other Western European countries. Although the preliminary agenda was set in the 1990s,[1] its actual political debate started only in 2011 with the submission and discussion of proposals to legalize the health care directive—also known as 'living will', which may be considered as a form of 'passive euthanasia'—which ultimately led to its unanimous approval in 2012 (Law 25/2012[2]). During this debate but mostly afterwards, the possible legalization of active euthanasia became a 'hot' topic in the political agenda in Portugal, particularly since April 2016, when a popular petition asking for the legalization of active euthanasia was submitted to Parliament.

Two opposing petitions were later extensively discussed in Parliament, and several bills were debated and voted on in six rounds of discussion. While the first of these rounds (2017–2018) led to the rejection of the four bills by a small margin, the following rounds—the second (2019–2021), third (late 2021), fourth (Mar. 2022–Jan. 2023), fifth (Mar.–Apr. 2023)—led to the approval of several bills from five different political parties and of the corresponding conciliatory texts, which were nevertheless rejected by the Constitutional Court or the President. Therefore, only after the sixth round of discussion, which took place in May 2023, was active euthanasia legalized in Portugal (Law 22/2023[3]).

As can be seen, this has been a particularly intense dispute, with several actors participating actively in the debates. Political parties were obvious active players, with the submission of several bills and their discussion in Parliament, in a quite heated debate. Several individual and organizational experts were also asked to participate in the debate both through hearings in meetings of the parliamentary committee and/or by issuing written expert opinions. Multiple interest groups also actively joined the debate, with strong participation through petitions and their presence in parliamentary hearings. Likewise, it is also important to highlight the great importance of both the President and the Constitutional Court, whose assessments of the bills led to four vetoes. Finally, the strong debate that also occurred outside Parliament should also be stressed, with intense discussion through opinion articles published in newspapers and television debates, as well as strong involvement from the top echelon of the Portuguese Catholic Church.

This widely participated political debate is therefore in line with what is expected to happen with the so-called morality issues, since it deals with deep core values but also because the policy itself tends not to be very complex, which enables its discussion by a wider set of actors (Heichel et al., 2013; Hurka et al., 2017; Knill, 2013). As demonstrated in the literature, the change (or stability) of policies related to morality issues depends on factors such as party politics and political cleavages, interest groups, religion and courts, as well as problem pressures, societal values and public opinion, or even international and transnational influences (Heichel et al., 2013; Knill, 2013; Schmitt et al., 2013). Likewise, the literature also suggests the importance to analyse the so-called veto actors, as well as the formation of eventual 'advocacy coalitions' (Schmitt et al., 2013; Mota & Fernandes, 2022). Finally, the literature also proposes that there is usually a difference in policy processes between countries from the so-called 'religious world'—where a religious cleavage in party-politics exists and therefore there is a higher chance for the politicization of these issues by political parties—and countries from the 'secular world'—where no such cleavage exists, meaning that there is a need for interest groups to be more active in the agenda-setting and formulation processes (Engeli et al., 2012b).

The legalization of active euthanasia in Portugal, which became only the eighth country in the world to do so—after the Netherlands, Belgium (both since 2002), Luxembourg (since 2009), Colombia (since 2014), Canada (since 2016), Spain and New Zealand (both since 2021) and six Australian states—may therefore come as a big surprise. On the one hand, Portugal still reveals comparatively high levels of religiosity (Moniz, 2018)—according to data from the European Values Study,[4] in 2017 62.15% of the population still considered religion to be important, one of the highest percentages in Western Europe. Likewise, the Portuguese population still tends to be rather conservative—data from the European Values Study also reveals that in 2017, only a small percentage of respondents considered that the acts of taking marijuana or hashish (13.86%), abortion (38.91%), homosexuality (41.98%) or euthanasia (42.89%) were justifiable.

This legislative outcome may, however, come as no surprise if one considers the recent 'wave of permissiveness' that Portugal has experienced since 2000—some examples are the decriminalization of drug use in 2001, the decriminalization of abortion in 2007, the legalization of same-sex marriage in 2010 and of the adoption by same-sex couples in 2016, the approval of self-determination of gender identity in 2018 or the approval of surrogacy in 2021. Portugal even placed first in the World Index of Moral Freedom of 2022, an index that measures individual citizens' freedom, with indicators regarding religion, bioethics, drugs, sexuality and gender (Álvarez et al., 2022).

Portugal is therefore a puzzling case study regarding morality issues in general, and euthanasia in particular. As mentioned before, most of the Portuguese population revealed conservative opinions regarding several morality issues in 2017, but the country has quite progressive regulations on these issues. What accounts for such outcomes? The few existing studies about the legislative advancements in morality issues—even if some of these do not use this theoretical lens—reveal that the discussion of abortion and issues about the rights of LGBTIQ people have also been widely politicized and have involved interest groups, such as feminist movements and LGBTIQ organizations (Alves et al., 2009; Santos, A., 2018; Santos, A. C., 2018; Mota & Fernandes, 2022). Moreover, Meyer-Resende and Hennig (2015) revealed that, unlike other Catholic churches, the Portuguese Catholic church did not participate actively in the discussion of these two issues. One may then wonder if these same trends were also relevant in the discussion on active euthanasia.

The debate on the possible legalization of active euthanasia held in Portugal between 2016 and 2023 is therefore the main object of study of this book, with a methodological strategy that will be further described in the following section.

1.2 Methodological Strategy

The goal of this book is to analyse the political and public debate on euthanasia in Portugal, to understand which the most active political and social actors in the debate were, either advocating for the decriminalization of euthanasia or against it. Moreover, we aim to understand if the arguments and frames used to support their positions are the same as those used in other countries, and if they were mainly 'rationally-oriented' or 'morally-oriented' (see Burlone & Richmond, 2018 and Sect. 2.2.2 about this distinction).

More specifically, the aim of this book is to study the participation of the following groups of actors in the debate on the issue of euthanasia held in Portugal: (i) political parties; (ii) veto players, such as the President and the Constitutional Court; (iii) experts; (iv) interest groups; (v) religious organizations, especially the Portuguese Catholic Church and (vi) other actors, particularly those who decided to write opinion articles in newspapers.

To this end, we will use the following sources of information: political documents, such as the two petitions submitted to Parliament, the bills presented throughout the various legislatures; minutes of the plenary parliament debates; the decisions from the Constitutional Court and the President; hearings held with interest groups and experts; the written opinions issued by organizational experts; official Catholic Church documents; opinion articles published in one mainstream newspaper and seven interviews held with political and social actors.

Concerning political documents, we analysed the content of the two petitions submitted to Parliament in 2016 and 2017, as well as the content of the 13 individual bills that were proposed by five political parties in rounds 1, 2 and 4, and of the five conciliatory texts submitted in rounds 2, 3, 4, 5 and 6. This analysis was mostly descriptive, identifying the details of the proposed regulation of active euthanasia.

Regarding the plenary parliament debates, we analysed the minutes of the plenary debates during which the bills (first, second and fourth rounds) or the conciliatory text (third and sixth rounds) were discussed, namely: May 29, 2018; February 20, 2020; November 5, 2021; June 9, 2022; March 31, 2023; and May 12, 2023. The analysis was focused on the arguments used to support or oppose the bills and the legalization of euthanasia. We opted to analyse these debates, rather than the meetings in committees, because the former tends to be more heated, and the differences in positions between political parties tend to be exacerbated. It is also important to mention that the debates became shorter and less tense throughout the six rounds, with fewer arguments being raised. This explains why our analysis is denser for the first two rounds.

We also analysed 33 hearings or written statements, 13 of them from organizational experts and the remaining 20 from civil society organizations and movements. The analysis of these hearings was also focused on the arguments they used to support or oppose the bills and the legalization of euthanasia.

Concerning documents from the Catholic Church, we proceeded with a search on the website of the Portuguese Episcopal Conference with the keywords 'euthanasia' and 'medically assisted death' and analysed all the documents identified.

Regarding opinion articles in written media, we chose to analyse opinion articles published in the newspaper *Público* between January 2016—when the first petition started to become more visible even before the submission to Parliament—and May 2023, when the final version of the *Law* was published. This newspaper was chosen because this is one of the most highly reputed daily newspapers in Portugal and is not directly connected to any specific political party. The analysed 169 opinion articles resulted from a search on this newspaper's website with the keywords 'euthanasia' and 'medically assisted death' and a selection of opinion articles that had euthanasia as their main topic and had a reflection on the issue itself and not only considerations on the legislative process.

Finally, we also analysed the content of seven interviews with political and social actors, namely representatives from political parties who voted in favour and against the legalization of euthanasia, as well as representatives from civil society movements on the two 'sides' of the debate. Although all relevant political parties were asked for an interview, only 5 (out of 10) answered our request. These interviews are mainly focused on the dynamics of coordination between actors, including among political parties and between the parties and civil society actors.

1.3   STRUCTURE OF THE BOOK

To complete the mentioned goals, this book will follow structure of eight chapters in addition to the current introductory Chap. 1.

Chapter 2 will present the main relevant literature review about the topics to be further empirically explored in the following chapters, namely the topics on policy formulation (Sect. 2.1) and morality issues and euthanasia (Sect. 2.2).

In Chap. 3 we start our empirical analysis by presenting the main generic characteristics of each of the six rounds of discussion, as well as a brief overview of the general contents of the final law.

Chapter 4 analyses the positions advocated by the 10 political parties that had parliament seats during the 6 rounds, as well as the used arguments to support those positions.

In Chap. 5 we analyse the content of in-person hearings and written opinions from 13 organizational experts, including actors often consulted by Parliament (5.1) and other actors which are not (Sect. 5.2).

Chapter 6 covers the analysis of the content of in-person hearings and written opinions from 20 interest groups, divided into 3 categories: the petitioners (Sect. 6.1), the interest groups dedicated to this cause (Sect. 6.2) and the religious interest groups (Sect. 6.3).

In Chap. 7 we analyse two fora of extra-parliamentary debate, including the discussion held by the hierarchy of the Portuguese Catholic Church (Sect. 7.1) and the opinion articles published in one newspaper (Sect. 7.2).

Chapter 8 analyses interviews with political and social actors about their perceptions of the relations among actors during this legislative process.

Finally, in Chap. 9 we present our conclusions and final considerations.

NOTES

1. According to a recent article in the Portuguese newspaper *Público*, the topic has been under discussion at least since 1995 and had one of its peaks during the discussion about the legalisation of the so-called 'living will' or 'health care directive' in 2011 and 2012. See https://www.publico.pt/2020/02/12/sociedade/noticia/eutanasia-portugal-ja-1995-tema-motivo-debate-1903701
2. The Law may be consulted in the following link: https://www.pgdlisboa.pt/leis/lei_mostra_articulado.php?nid=1765&tabela=leis&so_miolo=
3. The Law may be consulted in the following link: https://www.pgdlisboa.pt/leis/lei_mostra_articulado.php?nid=3648&tabela=leis&ficha=1&pagina=1&so_miolo=
4. See https://europeanvaluesstudy.eu/methodology-data-documentation/maps/

REFERENCES

Álvarez, G., Kotera, Y., & Pina, J. (2022). *World Index of Moral Freedom—WIMF 2022*. Foundation for the Advancement of Liberty.

Alves, M., Santos, A., Barradas, C., & Duarte, M. (2009). A despenalização do aborto em Portugal—Discursos, dinâmicas e acção coletiva: os referendos de 1998 e 2007 (The decriminalization of abortion in Portugal—Discourses, dynamics and collective action: The 1998 and 2007 referendums). *Oficina do CES no. 320.*

Burlone, N., & Richmond, R. (2018). Between morality and rationality: Framing end-of-life care policy through narratives. *Policy Sciences, 51,* 313–334.

Engeli, I., Green-Pedersen, C., & Larsen, L. (2012a). Introduction. In E. I. Engeli, C. Green-Pedersen, & L. T. Larsen (Eds.), *Morality politics in Western Europe: Parties, agendas and policy choices* (pp. 1–4). Palgrave Macmillan.

Engeli, I., Green-Pedersen, C., & Larsen, L. (2012b). The two worlds of morality politics—What have we learned? In E. I. Engeli, C. Green-Pedersen, L. T. Larsen, I. Engeli, C. Green-Pedersen, & L. T. Larsen (Eds.), *Morality politics in Western Europe* (pp. 185–199). Palgrave Macmillan.

Heichel, S., Knill, C., & Schmitt, S. (2013). Public policy meets morality: Conceptual and theoretical challenges in the analysis of morality policy change. *Journal of European Public Policy, 20*(3), 318–334. https://doi.org/10.1080/13501763.2013.761497

Hurka, S., Adam, C., & Knill, C. (2017). Is morality policy different? Testing sectoral and institutional explanations of policy change. *Policy Studies Journal, 45*(4), 688–712.

Knill, C. (2013). The study of morality policy: Analytical implications from a public policy perspective. *Journal of European Public Policy, 20*(3), 309–317. https://doi.org/10.1080/13501763.2013.761494

Meyer-Resende, M., & Hennig, A. (2015). Shunning direct intervention: Explaining the exceptional behaviour of the Portuguese Church hierarchy in morality politics. *New Diversities, 17*(1), 145–160.

Moniz, J. (2018). Índice de Religiosidade: Uma proposta de teorização e medição dos fenómenos religiosos contemporâneos (Religiosity Index: A proposal for theorizing and measuring contemporary religious phenomena). *Revista Brasileira de História das Religiões, 32,* 191–219.

Mota, L., & Fernandes, B. (2022). Debating the law of self-determination of gender identity in Portugal: Composition and dynamics of advocacy coalitions of political and civil society actors in the discussion of morality issues. *Social Politics: International Studies in Gender, State & Society, 29*(1), 50–70.

Preidel, C., & Knill, C. (2015). Euthanasia: Different moves towards punitive permissiveness. In E. C. Knill, C. Adam, & S. Hurka (Eds.), *On the road to permissiveness?: Change and convergence of moral regulation in Europe* (pp. 79–101). Oxford University Press.

Santos, A. (2018). A Institucionalização da Bioética e as Políticas Públicas de Saúde em Portugal (The Institutionalization of Bioethics and Public Health Policies in Portugal). Instituto Universitário de Lisboa, PhD thesis.

Santos, A. C. (2018). Luta LGBTQ em Portugal: duas décadas de histórias, memórias e resistências [LGBTQ struggle in Portugal: two decades of stories, memories and resistance]. *Transversos: Revista de História, 14,* 37–52.

Schmitt, S., Euchner, E.-M., & Preidel, C. (2013). Regulating prostitution and same-sex marriage in Italy and Spain: The interplay of political and societal veto players in two catholic societies. *Journal of European Public Policy, 20*(3), 425–441.

Weible, C., & Nohrstedt, D. (2012). The advocacy coalition framework: Coalitions, learning and policy change. In E. E. Araral Jr., S. Fritzen, M. Howlett, M. Ramesh, & X. Wu (Eds.), *Routledge handbook of public policy* (pp. 125–137). Routledge.

CHAPTER 2

Policy Formulation, Morality Issues and Euthanasia: A Literature Review

Abstract As mentioned in the introduction, this book analyses public decisions about the regulation of euthanasia in Portugal, which may be considered an example of a morality issue, depending on the use of morality-related policy frames or arguments during its discussion. According to the literature, morality issues may be considered a distinct type of public policy, particularly because its policy-making process has different characteristics, namely during the agenda-setting and formulation (sub-) processes.

Having in mind this context, this chapter discusses the relevant literature to understand our object of study—the process of policy formulation of regulation of the issue of active euthanasia in Portugal. This discussion is divided into two subchapters: the first one, which discusses the literature on policy formulation and the characteristics of the policy formulation actors in Portugal; the second one, which reviews the literature on morality issues, in general, and on euthanasia, in particular.

Keywords Euthanasia; Morality issues; Policy formulation; Advocacy coalitions; Portuguese political system

© The Author(s), under exclusive license to Springer Nature Switzerland AG 2023
I. S. Almeida, L. F. Mota, *Politics and Policies in the Debate on Euthanasia*, https://doi.org/10.1007/978-3-031-44588-0_2

2.1   Public Policy Formulation: The Main Important Actors in Portugal and Dynamics of Interaction Between Them

2.1.1   Policy Formulation

In the analysis of policy formulation, a preliminary concept that needs to be considered is that of public policies. As Knill and Tosun (2020) simply put it, public policies are groups of decisions which reflect courses of action or non-action which are taken by decision-makers (typically governments or parliaments) to deal with a particular issue. In this regard, it is also important to highlight that a public policy is created to mitigate/solve or prevent a public problem that needs attention, which means it is designed to meet a goal of public interest (Birkland, 2016, p. 12). This idea of public interest is nevertheless a slippery one, since different actors may have different conceptions about whether a certain issue is or is not a problem that requires public intervention or about the desirable interventions to deal with that issue (Birkland, 2016, p. 12).

The study of public policy must therefore cover not only the technical aspects of the prescribed interventions (the so-called policy instruments) or the produced outputs and outcomes of that intervention, but also the study of the political dimension, including the different policy actors involved throughout the policy-making process (polity) and the dynamics of interaction they develop with one another (politics) (Knill & Tosun, 2020; Howlett et al., 2020).

Although there are several theories and models related to the study of policymaking, the most famous heuristic in this field of study is that which divides the policy process into five stages, which are strongly related to a problem-solving approach: *(1) agenda-setting*, which includes the recognition of a certain issue as a problem and its preliminary discussion in the public, media and political spheres; *(2) formulation*, during which alternatives of intervention are drafted and discussed more technically; *(3) decision-making*, which implies the approval of a certain alternative by public authorities, which may imply change or maintaining the status quo; *(4) implementation*, during which the chosen policy instrument(s) is/are put in place and *(5) evaluation*, which implies the assessment of the implementation process, the results and the impacts (Jann & Wegrich, 2007).

As explained in the introduction, this book focuses particularly on the policy formulation stage of the regulation of active euthanasia in Portugal—although there is also a consideration of the policy agenda-setting stage since both stages are highly interrelated. This involves the process of drafting of proposals and the discussion of the different proposals (including the ones which advocate no change), as well as the different actors involved.

One important topic to take into consideration in this regard is that the identification and proposal of alternatives to public intervention is not a purely technical process, even if there is an increasing call for more rational decision-making processes based on scientific and technical evidence—the so-called evidence-based policymaking (Davies et al., 2000; Stone, 2007). First of all, there is a need to acknowledge there are several constraints (Howlett et al., 2020, pp. 135–138) to the formulation of proposals. These may be substantive constraints—those related to the policy issue itself, in the sense that the problem may be too complex and unsolvable (e.g. the elimination of homelessness by just providing a shelter, or the promotion of healthy eating habits by decree or they may be procedural constraints—those related to institutional features of the political system and the distribution of power between actors or constitutional rules (e.g. municipalities trying to formulate a policy on an issue which is under the control of Central Government, or the Parliament trying to legalize an issue that the Constitution forbids).

Likewise, there is a need to recognize that policy formulation is a highly political stage in which technical and political rationalities are intertwined, with the former being bigger or smaller depending on the degree of technicality of the issue at stake (Cohen et al., 1972; Lindblom, 1959). Indeed, it is in this stage of the policy-making process that political actors and interest groups tend to be more active (Knill & Tosun, 2020). It is unlike what happens during the agenda-setting stage, in which a more diverse group of actors may participate—the so-called policy universe, which comprises official actors (legislative, executive and judicial actors), as well as non-state actors such as citizens, interest groups, the media, think-tanks and political parties. Only a subset of these actors are usually active during policy formulation—the so-called policy subsystem (Howlett et al., 2020). The policy subsystem is, therefore, a theoretical construct that may comprise a very small group of policy actors, such as legislative committees, powerful interest groups and top bureaucrats—the so-called 'iron triangle' (Jordan, 1981)—or a wider group of actors, which in addition to the mentioned iron triangle also includes experts, opinion-makers, and specialized journalists—the 'issue networks' identified and named by Heclo (1978).

The diversity of policy actors involved in the policy formulation stage is indeed a matter of theoretical dispute and has given rise to different theoretical approaches. These include the pluralistic approach, which considers that a multitude of interest groups may have the capacity to influence the decision-making processes, as well as an elitist approach, which advocates that only a small set of powerful interest groups can actually influence decision-makers (Howlett et al., 2020, pp. 42–47). In this regard, a more recent theory uses the concept of policy networks to analyse the policy subsystem, arguing that multiple political and societal groups are often organized into clusters of interests (Howlett et al., 2020).

A rather important concept related to that of policy networks is advocacy coalitions, originally proposed by Sabatier (1988). According to this author, the policy subsystems of 'contested' issues—such as that of euthanasia—tend to be populated by at least two coalitions, which compete with one another in the defence of different ideals and interests based on the beliefs of their members. Moreover, it should be noted that each coalition tends to demonize the image and arguments of competing coalitions, based on ethical reasons, personal opinions or the beliefs they represent (Sabatier & Weible, 2007; Weible & Nohrstedt, 2012).

As explained in the introduction, we focus not only on the technical aspect of the policy proposals about the regulation of active euthanasia in Portugal but mostly on the political aspect of the dispute. Therefore, it is important to acknowledge the role that different policy actors may play in the agenda-setting and policy formulation stages. In this regard, it is relevant to make a distinction between official actors and non-official actors. In the first group, the following actors can be identified: *(a)* legislative actors, which apart from their role in producing new legislation, have also oversight roles; *(b)* executive actors, including the government and public sector organizations, which may have also legislative powers but are mostly devoted to implementing public policies and *(c)* judicial actors, which, depending on the judicial system, are mostly concerned with interpreting the law and applying it to particular cases but also have a very political role, namely when it comes to a constitutional review of legislation (Birkland, 2016, Ch. 4).

Regarding the non-official actors, another set of actors should also be considered (Birkland, 2016, Ch. 5), namely: *(d)* political parties, whose main roles are to debate ideas, formulate manifestos and prepare candidates, thus acting as a 'bridge' between 'worlds' of the politics and society; *(e)* citizens or 'the public', who may be involved in the policy-making process by voting or participating in interest groups, but may also be involved through information, consultation and co-production processes; *(f)* interest groups, which advocate for the interests their members care about, be they more self-interested or altruistic; *(g)* think tanks and experts, which are more concerned with producing knowledge to inform the decision-making process and *(h)* the media, which may act as agenda-setters of certain issues, as watchdogs of the political and economic elite, and as gatekeepers, since they may choose who is given 'voice'.

Taking into consideration the topic of the book, a description of the Portuguese political system is important, particularly regarding the public policymaking process.

2.1.2   Policy Formulation Actors in Portugal

Portugal is a republican unitary state, despite the existence of two autonomous regions—the Atlantic archipelagos of Madeira and Azores—and has had a democratic political system since the 'carnation' revolution of April 25, 1974, which ended an authoritarian regime that had been in power since 1926 (see Costa and Paris (2022) for further details).

After the revolution of 1974 and a subsequent turbulent period, a constitution was ratified in April 1976, which, in its original form, had a clear socialist orientation (Pinto & Paris, 2023). However, its seven revisions[1] somehow eroded this ideological alignment by enabling a market economy and the adhesion to the European Economic Union (currently the European Union) and to the Eurozone, as well as the promotion of a more democratic and solid political system.

The current version of the Portuguese Constitution of 1976 (CRP)[2] is a very extensive document with 196 articles, which, after a set of fundamental principles, is divided into four parts: Part I, dedicated to Fundamental Rights and Duties; Part II, which covers the Organisation of the Economy; Part III, focused on the Organisation of the Political Power and Part IV, dedicated to the Guaranteeing and Revision of the Constitution.

As regards the organization of political power, Portugal has a semi-presidential political system, although some consider this classification as controversial (Neto, 2023). Despite the importance of Presidential powers, we start this analysis by highlighting the roles of the unicameral

Parliament (also known as the Assembly of the Republic), which is composed of 230 members (MPs), who are elected in legislative elections for a term of four years based on party lists for each of the 22 constituencies. According to the Constitution, Parliament has several duties (arts. 161 to 170 of the CRP), among which are to produce legislation on various issues[3] and to oversee the Government's activities. Both activities take place in plenary meetings and meetings of specialized committees—currently, there are 14 of these committees, among which is the Committee on Constitutional Affairs, Rights, Freedoms and Guarantees, in which the debates about euthanasia have been taking place. In the words of Goes and Leston-Bandeira (2023, p. 143), these parliamentary committees have increasing power as they "… have been allowed to carry out research, ask for information, undertake parliamentary auditing sessions, and ask members of the public, officials and ministers to testify in committee hearings and inquiries (…) which in turn have been opened to media scrutiny".

It is important to remark that the balance of powers between Parliament and the Government is not determined by the former's formal powers but rather by the electoral results (see Table 1)—Parliament has weak initiative during majority governments and strong influence during minority governments (Goes & Leston-Bandeira, 2023, p. 140). In this regard, another important feature to acknowledge is the importance of parliamentary party groups, which dominate the Parliamentary agenda and debate and often impose discipline in the vote (Goes & Leston-Bandeira, 2023, pp. 141–142; Jalali & Teruel, 2019, p. 59). Dissent against party discipline and voting freedom for certain issues, particularly in the so-called 'matters of conscience' may nevertheless take place—particularly in the two main centre catch-all parties (see next paragraph) (Jalali & Teruel, 2019, p. 59).

Until 2015, the Portuguese Parliament only had six political parties[4]—the left-wing conservative Communist Party (PCP); the left-wing environmentalist Green Party (PEV), which always runs for elections in coalition with PCP; the left-wing liberal Left Bloc (BE); the centre-left Socialist Party (PS); the centre-right Social Democrat Party (PSD) and the right-wing Christian democrat and economically liberal Popular Party (CDS-PP). Since then, four other political parties have also become able to elect MPs: in 2015, the centre party People-Animal-Nature (PAN), which describes itself as being outside the left-right divide but is often described as being centre-left; and, in 2019, the left-wing liberal Free

Party (L), the right-wing Liberal Initiative (IL) and the extreme-right conservative and populist Enough! party (CH). In the following table, the distribution of seats in Parliament can be seen for recent decades, starting in 1995, which enables us to conclude that there are still two dominant centre political parties—PS and PSD (Fig. 2.1).

Concerning Parliament it is also important to mention the existence of a set of independent administrative entities,[5] among which the National Ethics Council for the Life Sciences, which, according to its website,[6] "(...) is an independent body created in 1990 to analyse (...) systematically the moral problems which arise out of scientific progress in the fields of biology, medicine or general health care". As mentioned by Santos, A. (2018), this is a very powerful actor on issues regarding bioethics.

Another very important political actor in Portugal is the Government, which, according to the Portuguese Constitution (art. 183 of the CRP), is composed of the Prime Minister, a potencial Deputy Prime Minister, Ministers, and Secretaries of State (i.e. Junior Ministers). Since October 2015, the Portuguese government has been led by António Costa, a

	Oct. 1995	Oct. 1999	Mar. 2002	Feb. 2005	Sep. 2009	Jun. 2011	Oct. 2015	Oct. 2019	Jan. 2022
PCP	13	15	10	12	13	14	15	10	6
PEV	2	2	2	2	2	2	2	2	0
BE	-	2	3	8	16	8	19	19	5
L	-	-	-	-	-	-	-	1 **	1
PS	112	115	96	121	97	74	86	108	120
PAN	-	-	-	-	-	0	1	4 **	1
PSD	88	81	105	75	81	108	84 *	79	77
IL	-	-	-	-	-	-	-	1	8
CDS-PP	15	15	14	12	21	24	18 *	5	0
CH	-	-	-	-	-	-	-	1	12
Parties in Government	PS (min.)	PS (min.)	PSD + PP (maj.)	PS (maj.)	PS (min.)	PSD + PP (maj.)	PS (min.) ***	PS (min.)	PS (maj.)

Fig. 2.1 Distribution of MPs in the legislative elections in Portugal (1995–2022). (* PSD and CDS-PP ran for office in a coalition named Portugal to the Front (PàF), but the elected MPs were still seated in two different party groups. ** PAN and L both lost one MP during the legislature, who remained in Parliament as non-aligned MPs. *** The socialist government had a formal and stable parliamentary agreement with the PCP, PEV and BE)

'professional politician', who was previously minister of several portfolios and Mayor of Lisbon. During its first term, his government counted on the parliamentary support of three other left-wing political parties—PCP, PEV and BE—while in the second term, it had no such support, which led to early elections. The third and current term has a parliamentary majority.

Also part of the executive power, public sector organizations tend to have low levels of autonomy vis-à-vis the Government, as they largely depend on it for their budget but also because top public sector executives still tend to be selected based on political proximity to the government rather than just meritocratic aspects, even if less so than in the past (Silva, 2020; Tavares, 2019). This is particularly true for the direct (general directorates, secretaries and inspectorates) and indirect administration (agencies and public enterprises), but less so for the so-called autonomous administration, composed of municipalities, autonomous regions and some professional orders.

Courts in general, and the Constitutional Court, in particular, are also very important actors in the policy-making process. The Portuguese Constitution determines, for instance, that courts are sovereign entities, and they are independent regarding other actors (arts. 202 and 203 of the CRP). Concerning the Constitutional Court, this document stipulates that its major role is to perform reviews on the constitutionality and legality (art. 223 of the CRP) of bills or already approved legislation when asked by other actors. These reviews may have a preventive nature and target bills upon a request submitted by the President, the Prime Minister or one-fifth of MPs (art. 278 of the CRP), or they may have a successive nature and target already approved norms upon a request submitted by the President, the President of the Assembly of Republic, the Prime Minister, the Ombudsman, the Prosecutor General or one-tenth of MPs (art. 281 of the CRP). These requests may be made regarding the entire document or targeting specific norms. In this regard, it is important to mention that if the Constitutional Court declares a bill unconstitutional or illegal, the President must veto it and return it to the legislating entity (nr 1 of art. 279 of the CRP). After that, that bill may only be enacted if the unconstitutional or illegal section(s) is expunged or confirmed by two-thirds of the legislating entity (nr 2 of the same article). If the bill is reformulated, the Constitutional Court may be again asked to review it (nr 3 of art. 279). Some research considers that the operation of the Portuguese Constitutional Court is in line with the idea of judicial politics since 10 (out of the 13) judges of the court are appointed by

Parliament for a non-renewable period of 9 years and they tend to demonstrate party loyalty (Garoupa & Tiede, 2023).

Finally, concerning the official actors, it is also important to acknowledge the role played by the President, who is the formal head of the Portuguese State and has several political functions related to other entities, such as the power to dissolve Parliament—which already happened once, in 2005—and to nominate the Prime Ministers, based on the results of the legislative elections, and to exonerate them (art. 133 of the CRP). Regarding the public policymaking process, the President has the task of analysing new bills and promulgating them, sending them to the Constitutional Court for a preventive review of the constitutionality or vetoing them (political veto) (art. 134 of the CRP). The President may also submit issues of national interest to referendum, after a proposal submitted by Parliament or the Government (arts. 115 and 134 of the CRP). Given these duties and the fact that the President runs for office as a citizen and not as a member of a political party, although political parties express their public support for candidates, there is no scholarly agreement on whether the Portuguese political system has a situation of cohabitation—when the President and the Prime-Minister may dispute power, particularly if they are from different political parties—or if the President "exercises an above-party moderating power" (Neto, 2023, p. 122). Since early 2016, the President is Marcelo Rebelo de Sousa, a now-retired university lecturer in constitutional law, who was a leader of the right-wing PSD in the late 1990s and was a famous political commentator on TV for more than a decade before his candidacy.

In addition to official actors, the importance of non-official actors in the Portuguese political system should also be acknowledged. Political parties are one important actor, as described earlier. An important feature of the party system is the continued dominance of two centrist catch-all parties—PS and PSD—which have led all the governments since 1976. Moreover, it is also important to highlight a significant institutionalization and consolidation of the party system and a certain ideological stability of these parties, particularly since 1995, but only a moderate polarization (Pratas & Bizzarro, 2023).

Interest groups are also other important actors to consider, even if lobbying is not regulated in Portugal and there is no tradition of strong involvement of civil society organizations in political life, except in the period immediately after the revolution of 1974. This is partly explained by the fact that most non-governmental organizations are more focused on service provision and recreational tasks than on advocacy functions (Franco, 2015). Nevertheless, it is important to acknowledge the importance that unions and employer organizations still have in negotiating labour policies, as well as of other interest groups which became more autonomous regarding political parties, even if they continue to nurture (informal) contacts with policymakers (Lisi & Loureiro, 2023). As Lisi and Marquez (2019) mention, these actors tend to prefer to directly approach politicians—either from government parties, particularly during majority governments, or opposition parties, namely during minority government—than to mobilize citizens to indirectly influence politicians. These contacts may take place through formal channels, such as petitions, meetings with MPs, hearings in committees and the submission of reports and information at the request of MPs, but also through informal channels, such as open letters, conferences, seminars, presence in TV or the publication of opinion articles in newspapers (Lisi & Marquez, 2019).

Taking into consideration our object of study, two types of interest groups should be highlighted. On the one hand, the professional orders, particularly those of physicians and lawyers, tend to be very powerful actors, particularly regarding issues related to the regulation of their profession but are also active in the discussion of policy issues from health care and justice sectors (Escada & Lucas, 2019). And, on the other hand, the Portuguese Catholic Church has had a more reserved attitude regarding the political arena due to a 'pact' during the democratic transition and a liberal orientation of the Cardinals António Ribeiro (in office between 1971 and 1998) and José Policarpo (in office between 1998 and 2013) (Meyer Resende, 2023). The current Cardinal Manuel Clemente decided nevertheless to take a more active role in the discussion of political issues, particularly those of a morality nature (Meyer-Resende & Hennig, 2015).

Another important non-official actor to consider is citizens. As mentioned in the Constitution, citizens in Portugal have the right to vote (art. 49 CRP), access public office (art. 50 CRP), be part of political associations and parties (art. 51 CRP) and submit petitions[7] (art. 52 CRP). Despite these rights, Cancela (2023) demonstrates that participation in elections has been progressively decreasing—for instance, the turnout rate of the 2022 legislative elections was only 52.2% and in 2019 only 48.57%, while the scenario is more or less the same or even worse for presidential elections and elections for the European Parliament. On the other hand, the level of citizen involvement in non-electoral political participation modes is increasing (particularly since 2012, the peak of the economic crisis) (Accornero & Pinto, 2023) but it is still rather low—the results of the European Social Survey about what respondents have done in the previous year demonstrate that only around 20% of the population have contacted politicians or been involved in petitions, while only around 10% have participated in protests or boycotts, and only around 5% have worked for a political party or other political groups (Magalhaes, 2023).

Finally, it is important to highlight that the mass media tends not to have a very adversarial approach but rather a 'sacerdotal' one, since, due to the generic shortage of investigative journalism, journalists largely depend on politicians as sources of information and as opinion-makers (Salgado, 2023).

2.2   Euthanasia as a Morality Issue

2.2.1   *Morality Issues: Concept, Types and Factors Explaining Policy Change and Stability*

Morality issues can be defined as issues whose political debate is focused on arguments related to moral, ethical and religious values, rather than material values (Heichel et al., 2013; Hurka et al., 2017; Knill, 2013). It is important to note that the regulation of these issues addresses decisions about the validity of a set of moral values and, therefore, their discussion often generate social conflicts over what is 'right' or 'wrong' (Budde et al., 2017).

According to some authors (e.g. Knill, 2013), four types of issues may fall into this category: *(a) 'life and death' issues*, such as abortion, euthanasia or artificial insemination; *(b) 'sexual behaviour' issues*, including prostitution, pornography and homosexuality; *(c) issues of 'addictive behaviours'*, such as gambling and drug use and *(d) issues of 'individual freedom and collective values'*, such as gun ownership. These issues may not be nevertheless directly considered moral issues, but only if the debates about them involve the significant use of moral frames and arguments. According to Knill (2013), if the debate is centred around moral values, the issue may be considered as an issue of 'manifest morality', and when the debate mobilizes other kinds of frames (e.g. financial/tax frames, when discussing prostitution or drug use), it constitutes an issue of 'latent morality'.

As explained in the introduction, morality issues in Europe have changed significantly in Europe in recent decades, which has led to increasing attention from scholars, who were able to identify a set of factors that explain patterns of policy change and stability, namely:

- *Problem pressure*, such as that caused by moral shocks or focus events that, by their projection or magnitude, may influence public opinion and the political agenda (e.g. a school shooting leading to the discussion of gun control);
- *Values of society and public opinion*, since the dominant values of society and public opinion may influence political parties, which try to follow the opinion of the majority of their voters to maximize electoral gains;
- *Religion*, not only because citizens' opinion may be influenced by their level of religiosity, but also because religious actors may act as interest groups, either directly addressing politicians or indirectly addressing (religious) citizens;
- *Party politics*, particularly when the national party system has relatively strong confessional/conservative parties and these have a direct cleavage with secular parties, which may lead secular/liberal parties to set these issues in the agenda, so as to have political gains in increasingly secularizing societies;
- *Interest groups*, which may try to influence politicians, either directly or indirectly through citizens, to set these issues in their agendas and to make decisions (of policy change or stability) aligned with the interests they advocate;
- *National and international courts*, since decisions of common courts, especially in countries of common law tradition, or of higher national or international courts (e.g. decisions of the European Court of Human Rights) may lead to the agenda-setting of morality issues or block proposals, as Constitutional Courts may act as veto players; and
- *International and transnational influences*, including international law and norms and emulation and diffusion dynamics.

In addition to identifying of these main explanatory factors of change and stability in morality policies, recent literature also identified patterns of change in different (groups of) countries. In this context, some studies identify differences between the so-called religious and secular 'worlds', which are distinguished by the existence or absence of a religious cleavage in the national party system (Engeli et al., 2012b; Heichel et al., 2013).

On the one hand, in the 'religious world', party disputes tend to have a greater influence on issues related to morality issues, justified by the religious cleavage between parties (Engeli et al., 2012b). For this reason, secular parties tend to politicize issues related to morality issues, creating a divergence with confessional parties, to 'embarrass' them in increasingly secular societies (Engeli et al., 2012b; Heichel et al., 2013).

On the other hand, in the 'secular world', in the absence of a religious cleavage between political parties, morality issues tend not to enter the political agenda so easily, to the extent that these are seen as undisputed issues and thus do not gain the same impact and/or conflict as in the 'religious world' (Engeli et al., 2012b; Heichel et al., 2013). Given this situation, social actors (e.g. civil society organizations) are primarily responsible for placing these issues on the agenda through advocacy activities (Knill, 2013).

In the comparative literature on morality issues, Portugal has often been considered as being part of the 'religious world' (Engeli et al., 2012b; Studlar et al., 2013, 2018). As Engeli et al. (2012b) explain, Portugal is a country with a much more recent process of secularization and where conservative parties—such as the Christian democrat party CDS-PP and the centre-right PSD—thought it would be "attractive to adopt 'nonsecular' platforms to attract confessional voters—that is, a religiously inspired platform, but without direct references to religion or church" (Engeli et al., 2012a). Moreover, although the religious-secular conflict had been deactivated after democratization as a strategy to consolidate democracy, the left-wing parties activated morality issues at the beginning of the twenty-first century, namely with the issue of abortion and the request for a second referendum, which led to the decriminalization of abortion in 2007 (Engeli et al., 2012b).

On the other hand, some scholars consider Portugal to be part of the 'secular world' as there is a very low share of MPs from religious parties, considering just the CDS-PP (Euchner, 2019, pp. 25, 39). Likewise, it is important to acknowledge the comparatively lesser involvement of the Portuguese Catholic church in politics, at least until 2013, for the reasons explained earlier in the previous section (Meyer-Resende & Hennig, 2015).

In this book we take Engeli et al.'s approach by considering Portugal as part of the 'religious world' since a significant part of the PSD—including the President of the Republic, who is a former leader of this party and was elected with its support—still advocates traditional religious values. Moreover, the party CH—which elected its first MP in 2019 but is now the third political party in Parliament, with 12 MPs—is a self-proclaimed religious party. Moreover, Lago (2023) explains that religion (namely church attendance) has become a strong predictor of voting behaviour in Portugal since 2005, unlike what had happened in the past, when the Socialist Party started to take positions against the Catholic church's views, such as on abortion, while the left-wing parties PCP and BE were always secular. Despite the importance that party cleavages may have had in the activation of morality issues in Portugal, recent studies also demonstrate the importance of the action from civil society movements and organizations in the agenda-setting of issues such as abortion or the rights of LGBTIQ people (Alves et al., 2009; Santos, A. C., 2018; Mota & Fernandes, 2022).

This scenario may thus help to explain the 'wave of permissiveness' that Portugal has experienced since 2000 regarding morality issues—some examples being the decriminalization of drug use in 2001, the decriminalization of abortion in 2007, the legalization of same-sex marriage in 2010 and of adoption by same-sex couples in 2016, the approval of self-determination of gender identity in 2018 or the approval of surrogacy in 2021.

The political debate on euthanasia which started in 2016 may then be another example of this activation of morality issues.

2.2.2   Euthanasia: Types of Regulation and Main Arguments Against and in Favour

The concept of euthanasia is difficult to define, as it may have different understandings for different people. As mentioned by Keown (2018, p. 9), the word is derived from the Greek (*eu* + *thanatos*) which means "a gentle and easy death". However, the use of this word by the ancient Greeks did not have the same meaning as it does nowadays (Lindsay, 2019).

As Keown (2018, p. 9) mentions, there are some common features in most, if not all, definitions of euthanasia: i) It involves a decision that has the effect of shortening of someone's life; ii) the process of dying is performed in a medical context and iii) the death is meant to benefit the patient, who usually is undergoing grave suffering. In this regard, it is important to stress that the notion of suffering includes not only physical suffering but also mental suffering, as well as situations in which the patient has extremely diminished capacities, even if without physical or mental suffering (Lindsay, 2019, p. 1).

Despite the mentioned generic agreement on that broader concept, there are different meanings attached to the term euthanasia. On the one hand, there is a need to distinguish voluntary, non-voluntary and involuntary acts of euthanasia. While voluntary euthanasia is an act following the patient's request, non-voluntary euthanasia happens when the request is performed by another person (e.g. a patient's relative) since the patient is unable to express their desires (either because of their health conditions or because they are babies or adults with dementia), and involuntary euthanasia occurs when the act is contrary to the patient's desires (Keown, 2018, p. 15; Lindsay, 2019, p. 2). In this regard, it is important to mention that campaigners for the legalization of euthanasia (or the reduction of penalties attached to it) argue they are advocating only for voluntary euthanasia (Keown, 2018, p. 15), which is also the idea we are using during this book unless expressly mentioned otherwise.

On the other hand, it is necessary to distinguish the acts of passive and active euthanasia and the concept of assisted suicide. Whereas passive euthanasia happens when a life-sustaining treatment is suspended with the understanding that it will probably cause death—an omission—active euthanasia implies an intentional act that will directly result in the patient's death (e.g. the injection of a lethal drug)—an action (Lindsay, 2019, p. 2). Assisted suicide happens when patients themselves administer the substance that will cause their death rather than a health professional (Preidel & Knill, 2015, p. 84). These circumstances thus help to explain why euthanasia is also known as "medically assisted death" or "physician-assisted death", terms we have opted not to use.

Other important concepts in this regard are also those of dysthanasia, which is a term for "(…) the prolongation of the death process, through treatment that only prolongs the biological life of the patient, without the quality of life and dignity [which] (…) can also be called therapeutic obstinacy" and of orthotanasia, which gives doctors the possibility (or even the duty) of not carrying out dysthanasia practices (Félix et al., 2013, p. 2734). Moreover, the discussion of euthanasia is also intertwined with the discussion about palliative care, which is a branch of medicine that is focused on improving the quality of life for patients who have life-threatening illnesses (Fontalis et al., 2018, p. 407).

As several authors mention, euthanasia is a much debated but very divisive issue, which brings to the table several philosophical, ethical, legal, political, and religious arguments, both in favour of and against its decriminalization (Ball, 2017; Keown, 2018; Lindsay, 2019; Preidel & Knill, 2015).

On the one hand, one of the main arguments against euthanasia is the inviolability/sanctity of life, which is related to the often-mentioned 'right to life', that is, the right not to be deliberately killed by someone else or by ourselves (Keown, 2018, p. 38). According to those who use this argument, any act to shorten patients' lives is intrinsically wrong (Keown, 2018, p. 40; Lindsay, 2019, p. 6). As Keown (2018, p. 38) explains, this principle is widely spread in Western thought due to the Judaeo-Christian tradition, which is based on the idea that "(…) human life is created in the image of God and is, therefore, possessed of an intrinsic dignity which entitles it to protection from unjust attack". This is therefore the main reason why religion-related organizations and groups are against euthanasia (Ball, 2017). Despite its religious origins, this principle is also used in secular terms, as it has influenced several criminal law codes and medical ethics, as well as important international instruments of law such as the United Nations Universal Declaration of Human Rights (Keown, 2018, p. 49).

Another common argument against euthanasia is that it contravenes the purpose of medicine and medical ethics. As Ball (2017, p. 34) mentions, the Hippocratic Oath and other medical ethics codes stress that the role of the physicians is to heal, rather than to kill. Therefore, the practice of euthanasia by physicians would be a breach of medical ethics and would lead to decreased trust in physicians by citizens (Lindsay, 2019, p. 10), fearing that they will not be presented with all possible solutions or that euthanasia can be presented as the only possible path.

A third argument against euthanasia is that the potential for abuse, in the sense that vulnerable social groups (such as people with disability, elderly people, minorities and poor people), or those who think they may be a burden to their relatives and caregivers, may be pushed to ask for euthanasia (Ball, 2017, p. 35). This argument is particularly valid in jurisdictions with a lack of sufficient and good health and welfare services, particularly those related to palliative and elderly care. As Lindsay (2019, p. 9) mentions, there is a fear that some patients may be pressured to ask for euthanasia or may not be given enough information, or, even worse, physicians may start to euthanize people without a clear request. As Keown (2018) mentions, several studies demonstrate that healthcare professionals are not being strict on the criteria for performing acts of euthanasia and are not registering them properly in the Netherlands and Belgium. In this regard, Lindsay (2019, p. 9) also stresses the existence of studies which conclude that around 25% of cases of euthanasia in the Netherlands and Belgium take place without a clear prior or contemporaneous request.

Finally, a fourth argument against euthanasia, and probably one of the most common ones, is the fear of the slippery slope, that is the fear of evolution or relaxation of the criteria for euthanasia up to the point of non-voluntary or involuntary euthanasia starting to take place (Ball, 2017, p. 36). Opponents of euthanasia give the example of the later permission of 'minors' to be euthanized in Belgium or the (so far unsuccessful) bill in the Netherlands which tried to give the option of euthanasia to elderly people who were "tired of life" (Lindsay, 2019, p. 11; Keown, 2018).

On the other hand, the main argument advocated in favour of euthanasia is that of the need to respect patients' autonomy and individual liberty (or self-determination) to decide when and how they want to die (Ball, 2017, p. 20; Fontalis et al., 2018). As Fontalis et al. (2018, p. 408) mentions, autonomy is also deeply related to the capacity of competent adults to decide about the medical care they want to receive based on information provided by healthcare professionals.

Another very common argument in favour of legalizing euthanasia is that of compassion since the act of euthanasia helps patients to end their suffering or condition, thus being a way to 'die with dignity' and an act of mercy (Ball, 2017, p. 21; Fontalis et al., 2018; Inbadas et al., 2017). Campaigners for the legalization of euthanasia add that human life should only be considered 'sacred' if it contributes to happiness, health and joy.

A third argument in favour of legalizing euthanasia is related to the argument opposing the slippery slope, which argues that the regulations that approved active euthanasia or assisted suicide have rigorous protections and criteria that prevent situations of non-voluntary or involuntary acts of euthanasia from taking place (Ball, 2017, p. 22).

Finally, another argument in favour of legalizing euthanasia is that of consistency, since some consider there is no great difference between stopping medical treatments (passive euthanasia)—which is legal in several countries—and administering some lethal drug that helps patients to die, as the outcome is the same—the patient's death (Lindsay, 2019, p. 5).

As can be seen, several arguments have been raised in favour or against the legalization of euthanasia, some of which have a more moral or rational nature. As Burlone and Richmond (2018) mention, those about respect for individual autonomy and the idea of 'dying with dignity' (in favour) or those about the inviolability/sanctity of life and medical ethics (against) may be considered moral arguments. On the other hand, those of ending insufferable pain and suffering (in favour) or the fear of the slippery slope and the lack of palliative care are arguments of a rational nature (Burlone & Richmond, 2018).

Given what has been presented, it is no surprise that this topic is particularly divisive. However, unlike other morality issues, permissiveness in the regulation of euthanasia has been relatively steady in most countries, even if its philosophical discussion has earlier roots. As Lindsay states, although the term euthanasia has Greek origins and there are some discussions about it in ancient Greek writings, the deeper philosophical discussion on euthanasia dates to the late nineteenth and early twentieth centuries when a discussion about eventual legislative reforms started to be discussed in the United States and the United Kingdom (Ball, 2017, p. 7; Lindsay, 2019). The debates about euthanasia during those earlier periods were nevertheless somehow different from today's discussions, as they were more focused on whether euthanasia fitted the "collective welfare" rather than on being about individual autonomy (Ball, 2017, p. 7). As mentioned by Ball (2017, p. 8), the discussion of euthanasia was deeply connected with the eugenics movement, which advocated the removal from society of 'undesirable'—that is, " 'imbeciles', feeble-minded, criminals, and other unfit persons", either through sterilization or even euthanasia. This discussion ultimately led to the approval of quite radical bills on euthanasia in two Midwestern states of the United States—Iowa and Ohio—which then included acts of ending the lives of patients with incurable diseases and extremely injured people but also involuntary deaths of "the physically disabled, the mentally handicapped, 'lunatics and idiots' and other state dependents" (Ball, 2017, pp. 9–10). The debate around these bills was nevertheless very intense and ultimately led to their being rejected, even thought the discussion continued for several decades (Ball, 2017, pp. 11–12).

These radical bills and the association between euthanasia and the eugenics movement with some practices of the Nazi party may help to explain why the topic of euthanasia was dormant for most of the twentieth century. As Ball (2017, p. 48) highlights, the debate about euthanasia started to become more serious in the United States only in the 1980s when stronger pressure groups started to emerge. A similar pattern was also observed in Europe, where almost no change took place during the twentieth century (Preidel & Knill, 2015).

As Preidel and Knill (2015) demonstrate in their analysis, the evolution of the regulation of euthanasia in sixteen European countries over six decades (1960–2010) was very slow. As these authors explain, between 1960 and 1990 only five countries allowed some type of euthanasia: assisted suicide in Switzerland and Sweden and passive euthanasia in the Netherlands, Ireland and Spain (Preidel & Knill, 2015). This scenario slightly changed during the 1990s and the first of the twenty-first century, with several countries allowing some type of euthanasia: Germany, Finland, the United Kingdom, Austria, Norway, Israel, France and Denmark started to permit some form of passive euthanasia, along with Spain and Ireland that already permitted it; assisted suicide continued to be permitted in Switzerland and Sweden; and active euthanasia started to be legal in the Netherlands and Belgium (Preidel & Knill, 2015). The Netherlands and Belgium were indeed the first two countries in the world to legalize active euthanasia (both in 2002), followed, in the meantime, by Luxembourg (2009), Colombia (2014), Canada (2016), Spain (2021), New Zealand (2021), as well as by six Australian states. Nowadays, passive euthanasia is also permitted in most countries in Europe, South America and North America.

As can be seen, there has been a growing number of countries which permit some sort of euthanasia. As Preidel and Knill (2015, p. 80) assert, this evolution is mostly related to a change towards societal individualization, that is, a growing importance attributed to individual autonomy, as well as to the development of healthcare conditions and new health technologies, which permit higher life expectancy and enable "people to be kept alive artificially".

As the previously presented literature on morality issues reveals, the evolution of the regulation of euthanasia is also related to its public acceptance. As can be seen in Fig. 2.2, the results of the different rounds of the European Values Study (1981–2017) in 40 European countries[8] reveal that since the 1980s there has been an overall increase in the percentage of people who think euthanasia is a justifiable act. While in 1981 only two countries had at least nearly half of the population in favour of euthanasia—Denmark and the Netherlands—in 2017, eighteen European countries had the same situation—Denmark (73.14%), the Netherlands (71.86%), Germany (69.91%), Finland (67.56%), France (63.13%), Sweden (61.13%), Iceland (62.85%), the U.K. (61.71%), Norway (59.31%), Slovenia (58.74%), Czechia (58.68%), Switzerland (57.65%), Austria (56.59%), Spain (56.51%), Estonia (55.91%) and Italy (53%), as

	1981	1990	1999	2008	2017	Δ 1999-2017	Δ 2008-2017
Albania	-	-	-	18.74	19.03		0.29
Armenia	-	-	-	17.38	19.04		1.66
Austria	-	29.88	37.43	38.92	56.59	19.16	17.67
Belgium	28.09	45.3	55.26	64.11	-		
Belarus	-	-	48.62	41.94	44.79	**-3.83**	2.85
Bosnia and Herz.	-	-	-	21.08	27.77		6.69
Bulgaria	-	28.23	30.49	36.49	30.51	0.02	**-5.98**
Cyprus	-	-	-	11.12	17.09		5.97
Croatia	-	-	31.31	33.26	37.4	6.09	4.14
Czechia	-	33.02	49.96	48.11	58.68	8.72	10.57
Denmark	56.19	51.45	62.36	64.4	73.14	10.78	8.74
Estonia	-	33.58	46.24	40.7	55.91	9.67	15.21
Finland	-	57.43	48.91	54.6	67.56	18.65	12.96
France	41.33	46.44	57.33	63.87	63.13	5.8	**-0.74**
Georgia	-	-	-	15.74	21.65		5.91
Germany	-	-	37.08	42.35	69.91	32.83	27.56
Hungary	-	40.09	31.46	38.28	42.95	11.49	4.67
Iceland	34.87	40.83	48.09	53.14	62.85	14.76	9.71
Ireland	12.45	17.37	24.75	31.76	-		
Italy	22.61	29.23	31.73	39.17	53	21.27	13.83
Latvia	-	35.38	47.09	41.71	-		
Lithuania	-	28.3	48.71	43.77	46.85	**-1.86**	3.08
Luxembourg	-	-	51.46	56.57	-		
Macedonia	-	-	-	23.66	18.65		**-5.01**
Montenegro	-	-	-	22.89	23.2		0.31
Netherlands	49.15	54.52	63.23	63.11	71.86	8.63	8.75
Norway	19.65	35.93	-	51.34	59.31		7.97
Poland	-	12.65	26.25	26.29	37.05	10.8	10.76
Portugal	-	25.88	27.76	41.19	42.89	15.13	1.7
Romania	-	27.85	25.41	24.4	23.31	**-2.1**	**-1.09**
Russia	-	-	51.09	37.32	42.64	**-8.45**	5.32
Serbia	-	-	-	28.2	33.02		4.82
Slovakia	-	26.95	43.98	38.09	42.89	-1.09	4.8
Slovenia	-	35.19	48.59	50.11	58.74	10.15	8.63
Spain	23.42	31.96	41.4	56.44	56.51	15.11	0.07
Sweden	38.61	44.88	56.4	61.65	63.13	6.73	1.48
Switzerland	-	-	-	45.04	57.65		12.61
Turkey	-	-	19.73	13.42	16.97	**-2.76**	3.55
United Kingdom	38.08	41.39	44.26	51.56	61.71	17.45	10.15
Ukraine	-	-	46.66	29.83	40.5	**-6.16**	10.67

Fig. 2.2 Evolution of public acceptance of euthanasia in European countries (1981–2017). (Percentage of surveyed people who consider that euthanasia is justifiable, based on the following question: "Please tell me whether you think euthanasia (terminating the life of the incurably sick) can always be justified, never be justified, or something in between". Rated on a scale from 1 (never justified) to 10 (always justified). Source: European Values Study)

well as Belgium and Luxembourg, who had surpassed that level of acceptance in 2008. Unsurprisingly, these countries are mainly from so-called
Western Europe, while the results from Eastern Europe tend to have lower
levels of acceptance, which, in some cases, is even decreasing. These results
may be then related to the level of religiosity, as Cohen and his associates
(2014) demonstrate for the 2008 results.

This broader acceptance of euthanasia in several Western European
countries is nevertheless not in line with the opinion of physicians. As
Emanuel and his associates (2016) mention, several studies indicate that
the percentage of support for euthanasia is much lower among physicians
than among the public. In this regard, these authors stress the results of a
study developed in 2014 by Medscape, for which 21,531 physicians from
seven countries were surveyed, which indicate that "US physicians were
most supportive, with 54% agreeing, while a minority of physicians in
Germany (47%), United Kingdom(47%), Italy (42%), France (30%), and
Spain (36%) concurred that PAS [physician assisted suicide] should be
permitted" (Emanuel et al., 2016, p. 81).

As also shown in the figure, there has been an increase in public acceptance of euthanasia in Portugal. While only a quarter of the population
was in favour of legalizing euthanasia in 1990, almost half of the population was in favour in 2017, and this percentage is now much higher.
According to two polls performed in 2020[9] and 2023,[10] around two-thirds
of the Portuguese population is in favour of legalizing euthanasia, a level
of support that tends to be even higher among young people.

Concerning the acceptance of euthanasia by physicians in Portugal, only
one recent study was identified, which surveyed 251 physicians in 2017,
with 58.2% of them declaring their support for euthanasia (Silva et al.,
2019). However, this study also reveals that the level of support differs
greatly between age groups and groups of seniority: 25–45 y.o. (66.2%),
46–65 y.o. (47.8%) and over 65 y.o. (36.4%); 0–10 years of experience
(69.4%), 11–20 years (61.1%) and over 20 years (42.6%) (Silva et al., 2019).

As mentioned in the introduction, the topic of euthanasia entered the
political agenda in Portugal in 2011 when four parties proposed bills to
regulate passive euthanasia in what was named the 'living will' or 'health
care directive' leading to the approval of Law 25/2012. According to this
law, advanced care directives, in the form of living wills, are documents in
which a person, who is a legal adult and has mental capacities, may manifest in advance a conscious, free and informed will regarding the healthcare he/she wishes to receive (article 1). This document, which may be

changed or revoked at any time, is valid for five years after its signature, although it may be renewed (articles 7 and 8). Another important detail of the law is the right of conscious objection attributed to healthcare professionals (article 9).

As Fig. 2.3 reveals, the number of 'living wills' was very low before active euthanasia started to be discussed and the topic re-entered the public agenda in 2017. The number of living wills increased again in 2021, 2022 and 2023 when the debate about active euthanasia became more involved.

Finally, before we proceed with the analysis of the debate, it is important to highlight that, as in most countries, there was no specific prohibition of euthanasia in the Constitution or the legal system in Portugal.

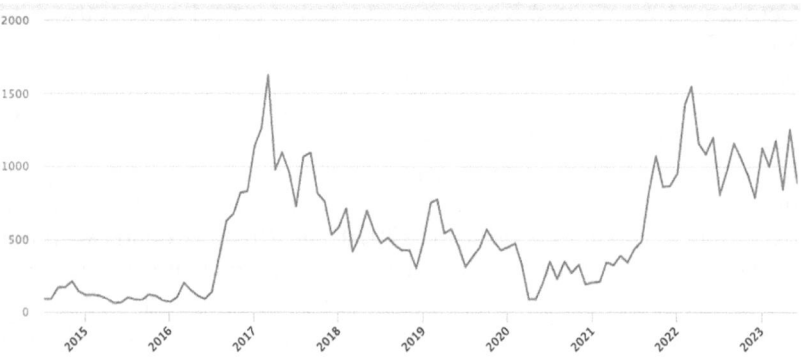

Fig. 2.3 Total number of active 'living wills' in Portugal. (Source: Transparência SNS website [https://transparencia.sns.gov.pt/explore/dataset/registo-de-testamentos-vitais/analyze/?flg=pt&disjunctive.ars&disjunctive.aces&sort=tempo&dataChart=eyJxdWVyaWVzIjpbeyJjaGFydHMiOlt7InR5cGUiOiJsaW5lIiwiZnVuYyI6IlNVTSIsInlBeGlzIjoidG90YWxfdGVzdGFtZW50b3Nfdml0YWlzX2RlX3V0ZW50ZXMiLCJjb2xvciI6IiM4ZGEwY2IiLCJjZ2llbRpZmljRGlzcGxheSI6dHJ1ZX1dLCJ4QXhpcyI6InRlbXBvIiwibWF4cG9pbnRzIjoiIiwid2ltZXNjYWxlIjoibW9udGgiLCJzb3J0Ijoiiwic2VyaWVzQnJlYWtkb3duIjoiIiwic2VyaWVzQnJlYWtkb3duVGltZXNjYWxlIjoiiwiY29uZmlnIjp7ImRhdGFzZXQiOiJyZWdpc3RvLWRlLXRlc3RhbWVudG9zLXZpdGFpcyIsIm9wdGlvbnMiOnsiZmxnIjoicHQiLCJkaXNqdW5jdGl2ZS5hcnMiOnRydWUsImRpc2p1bmN0aXZlLmFjZXMiOnRydWUsInNvcnQiOiJ0ZW1wbyJ9fX1dLCJkaXNwbGF5TGVnZW5kIjp0cnVlLCJhbGlnbk1vbnRoIjp0cnVlfQ%3D%3D])

GlzcGxheeGzVZVuZCI6dHJ1ZSwiYWxpZ25Nb250aCI6dHJ1ZX0=])

Nevertheless, it is also important to highlight that the Portuguese Constitution, in line with the United Nations Universal Declaration of Human Rights and the European Convention on Human Rights, specifies, in its article 24, that "human life is inviolable". As we will demonstrate, this principle was often mentioned by several actors, as well as the principle mentioned in article 26 of the Constitution which stipulates that "everyone is accorded the rights to personal identity, to the development of personality, to civil capacity [and] to citizenship (…)" or the principle expressed in article 41 that postulates that "The freedom of conscience, of religion and the form of worship is inviolable".

As concerns the legal system, there was no law specifically prohibiting euthanasia. However, article 134 of the Portuguese Penal Code[11], approved in 1995, specifies, in point number 1 that "whoever kills another person, determined by a serious, instant and express request that he or she has made, is punished by imprisonment for up to three years". In the same vein, article 135 of the same code, specifies that "whoever incites another person to commit suicide, or provides help for that purpose, is punished with imprisonment for up to 3 years, if suicide is actually attempted or consummated".

The alteration of these two articles, and consequent approval of active euthanasia and assisted suicide, was precisely the purpose of the petition submitted in April 2016, which propelled a huge debate that lasted eight years, and culminated in the approval of Law 22/2023. The following chapters are dedicated to analysing the various aspects of this debate.

Notes

1. The seven revisions took place in 1982, 1989, 1992, 1997, 2001, 2004 and 2005. See https://www.pgdlisboa.pt/leis/lei_mostra_articulado.php?nid=4&tabela=leis or · https://www.parlamento.pt/RevisoesConstitucionais/Paginas/default.aspx

2. See https://www.parlamento.pt/Legislacao/Paginas/ConstituicaoRepublicaPortuguesa.aspx. or, for a version in English, see https://www.parlamento.pt/sites/EN/Parliament/Documents/Constitution7th.pdf

3. Some of these issues are exclusively legislated by the Parliament (mostly issues of political organization) (article 164 of the CRP), while other issues may also be legislated by the Government (mostly related to sectoral policies) (articles 165 and 198 of the CRP).

4. To further explore the history of the traditional and new political parties in Portugal, please consult Freire (2023) and Marchi and Alves (2023).

5. The full list of independent administrative entities can be consulted via the following link: https://www.parlamento.pt/sites/EN/Parliament/Paginas/Indepedent-administrative-bodies.aspx
6. See https://www.cnecv.pt/en/history
7. As of October 2020, 7500 signatures are required for a petition to be mandatorily discussed in Parliament, while previously only 4000 signatures were necessary.
8. We excluded countries without any results or with results for only one year.
9. See   https://sol.sapo.pt/artigo/686756/mais-de-70-dos-portugueses-sao-a-favor-da-eutanasia-mas-mais-de-80-querem-referendo
10. See https://observador.pt/2023/02/26/sondagem-61-dos-portugueses-votariam-a-favor-da-eutanasia/
11. See   https://www.pgdlisboa.pt/leis/lei_mostra_articulado.php?ficha=101&artigo_id=&nid=109&pagina=2&tabela=leis&nversao=&so_miolo=

REFERENCES

Accornero, G., & Pinto, P. (2023). Movements at the border. Conflict and protest in Portugal. In E. J. Fernandes, P. Magalhaes, & A. Pinto (Eds.), *The Oxford handbook of Portuguese politics* (pp. 457–471). Oxford University Press.

Alves, M., Santos, A., Barradas, C., & Duarte, M. (2009). A despenalização do aborto em Portugal—Discursos, dinâmicas e acção coletiva: os referendos de 1998 e 2007 (The decriminalization of abortion in Portugal—Discourses, dynamics and collective action: The 1998 and 2007 referendums). *Oficina do CES no. 320.*

Ball, H. (2017). *The right to die: A reference handbook.* ABC-CLIO.

Birkland, T. (2016). *An introduction to the policy process theories, concepts, and models of public policy making* (4th ed.). Routledge.

Budde, E., Knill, C., Fernández-i-Marín, X., & Preidel, C. (2017). A matter of timing: The religious factor and morality policies. *Governance, 31*(1), 45–63.

Burlone, N., & Richmond, R. (2018). Between morality and rationality: Framing end-of-life care policy through narratives. *Policy Sciences, 51*, 313–334.

Cancela, J. (2023). Electoral Turnout. In E. J. Fernandes, P. Magalhaes, & A. Pinto (Eds.), *The Oxford handbook of Portuguese politics* (pp. 291–307). Oxford University Press.

Cohen, J., Van Landeghem, P., Carpentier, N., & Deliens, L. (2014). Public acceptance of euthanasia in Europe: A survey study in 47 countries. *International Journal of Public Health, 59*, 143–156.

Cohen, M., March, J., & Olsen, J. (1972). A Garbage can framework of organizational choice. *Administrative Science Quarterly, 17*(1), 1–25.

Davies, H., Nutley, S., & Smith, P. (2000). *What works? Evidence-based policy and practice in public services.*

Emanuel, E., Onwuteaka-Philipsen, B., Urwin, J., & Cohen, J. (2016). Attitudes and practices of euthanasia and physician-assisted suicide in the United States, Canada, and Europe. *Clinical Review & Education, 36*(1), 79–90.

Engeli, I., Green-Pedersen, C., & Larsen, L. (2012a). Introduction. In E. I. Engeli, C. Green-Pedersen, & L. T. Larsen (Eds.), *Morality politics in Western Europe: Parties, agendas and policy choices* (pp. 1–4). Palgrave Macmillan.

Engeli, I., Green-Pedersen, C., & Larsen, L. (2012b). The two worlds of morality politics—What have we learned? In E. I. Engeli, C. Green-Pedersen, L. T. Larsen, I. Engeli, C. Green-Pedersen, & L. T. Larsen (Eds.), *Morality politics in Western Europe* (pp. 185–199). Palgrave Macmillan.

Escada, M., & Lucas, T. (2019). A desregulação profissional em Portugal durante a Troika: o caso da Ordem dos Médicos e dos Advogados (Professional deregulation in Portugal during the Troika: The case of Physicians and Lawyers Professional Associations). In E. M. Lisi & M. Lisi (Eds.), *Grupos de Interesse e Crise Económica em Portugal (Interest groups and economic crisis in Portugal)* (pp. 207–236). Edições Sílabo.

Euchner, E.-M. (2019). *Morality politics in a secular age: Strategic parties and divided governments in Europe.* Palgrave Macmillan.

Félix, Z., Costa, S., Alves, A., Andrade, C., Duarte, M., & Brito, F. (2013). Eutanásia, distanásia e ortotanásia: uma revisão integrativa da literatura (Euthanasia, dysthanasia and orthothanasia: An integrative literature review). *Ciência & Saúde Coletiva, 18*(9), 2733–2746. https://doi.org/10.1590/S1413-81232013000900029

Fontalis, A., Prousali, E., & Kulkarni, K. (2018). Euthanasia and assisted dying: What is the current position and what are the key arguments informing the debate? *Journal of the Royal Society of Medicine, 111*(11), 407–413.

Franco, R. (2015). *Diagnóstico das ONG em Portugal.* Fundação calouste Gulbenkian.

Freire, A. (2023). The centre-left and the radical left in Portuguese democracy, 1974–2021. In E. J. Fernandes, P. Magalhaes, & A. Pinto (Eds.), *The Oxford handbook of Portuguese politics* (pp. 88–101). Oxford University Press.

Garoupa, N., & Tiede, L. (2023). Judicial politics in Portugal. In E. J. Fernandes, P. Magalhaes, & A. Pinto (Eds.), *The Oxford handbook of Portuguese politics* (pp. 164–180). Oxford University Press.

Goes, E., & Leston-Bandeira, C. (2023). The role of the Portuguese parliament. In E. J. Fernandes, P. Magalhaes, & A. Pinto (Eds.), *The Oxford handbook of Portuguese politics* (pp. 136–148). Oxford University Press.

Heclo, H. (1978). Issue networks and the executive establishment. In E. A. King (Ed.), *The new American political system* (pp. 87–124). American Enterprise Institute Press.

Heichel, S., Knill, C., & Schmitt, S. (2013). Public policy meets morality: Conceptual and theoretical challenges in the analysis of morality policy change. *Journal of European Public Policy, 20*(3), 318–334. https://doi.org/10.108 0/13501763.2013.761497

Howlett, M., Ramesh, M., & Perl, A. (2020). *Studying public policy: Principles and processes*(4th ed.). Oxford University Press. Obtido de https://global.oup.com/ academic/product/studying-public-policy-9780199026142?q=Michael%20 Howlett&lang=en&cc=pt#

Hurka, S., Adam, C., & Knill, C. (2017). Is morality policy different? testing sectoral and institutional explanations of policy change. *Policy Studies Journal, 45*(4), 688–712.

Inbadas, H., Zaman, S., Whitelaw, S., & Clark, D. (2017). Declarations on euthanasia and assisted dying. *Death Studies, 41*(9), 574–584.

Jalali, C., & Teruel, J. (2019). Parliamentary party groups in the Iberian democracies. In E. J. Fernandes & C. Leston-Bandeira (Eds.), *The Iberian legislatures in comparative perspective* (pp. 49–70). Routledge.

Jann, W., & Wegrich, K. (2007). Theories of the policy process. In E. F. Fischer, G. Miller, & M. Sidney (Eds.), *Handbook of public policy analysis: Theory, politics, and methods* (pp. 43–62). CRC Press.

Jordan, A. (1981). Iron triangles, Woolly corporatism and elastic nets: Images of the policy process. *Journal of Public Policy, 1*(1), 95–124.

Keown, J. (2018). *Euthanasia, ethics and public policy—An arguments against legalisation* (2nd ed.). Cambridge University Press.

Knill, C. (2013). The study of morality policy: Analytical implications from a public policy perspective. *Journal of European Public Policy, 20*(3), 309–317. https://doi.org/10.1080/13501763.2013.761494

Knill, C., Adam, C., & Hurka, S. (2015). *On the road to permissiveness? Change and convergence of moral regulation in Europe*. Oxford University Press.

Knill, C., & Tosun, J. (2020). *Public policy: A new introduction* (2nd ed.). Palgrave Macmillan. Obtido de https://www.macmillanihe.com/page/detail/public-policy-christoph-knill/?k=9780230278387

Lago, I. (2023). Voting behaviour. In E. J. Fernandes, P. Magalhães, & A. Costa Pinto (Eds.), *The Oxford handbook of Portuguese politics* (pp. 276–290). Oxford University Press.

Lindblom, C. (1959). The science of muddling through. *Public Administration Review, 19*(2), 79–88.

Lindsay, R. (2019). Euthanasia. In E. H. LaFollette (Ed.), *International encyclopedia of ethics*. John Wiley & Sons Ltd.

Lisi, M., & Loureiro, J. (2023). Interest groups, business associations and unions. In E. J. Fernandes, P. Magalhaes, & A. Pinto (Eds.), *The Oxford handbook of Portuguese politics* (pp. 423–439). Oxford University Press.

Lisi, M., & Marquez, L. (2019). Interest groups in the Iberian parliaments. In E. J. Fernandes & C. Leston-Bandeira (Eds.), *Iberian legislatures in comparative perspective* (pp. 130–148). Routledge.

Magalhaes, P. (2023). Citizens and politics: Support and engagement. In E. J. Fernandes, P. Magalhaes, & A. Pinto (Eds.), *The Oxford handbook of Portuguese politics* (pp. 244–261). Oxford University Press.

Marchi, R., & Alves, A. (2023). The right and far-right in the Portuguese democracy (1974–2022). In E. J. Fernandes, P. Magalhaes, & A. Pinto (Eds.), *The Oxford handbook of Portuguese politics* (pp. 102–118). Oxford University Press.

Meyer Resende, M. (2023). The relations between the Catholic church and the political arena in Portugal. In E. J. Fernandes, P. Magalhaes, & A. Pinto (Eds.), *The Oxford handbook of Portuguese politics* (pp. 472–486). Oxford University Press.

Meyer-Resende, M., & Hennig, A. (2015). Shunning direct intervention: Explaining the exceptional behaviour of the Portuguese Church hierarchy in morality politics. *New Diversities, 17*(1), 145–160.

Mota, L., & Fernandes, B. (2022). Debating the law of self-determination of gender identity in Portugal: Composition and dynamics of advocacy coalitions of political and civil society actors in the discussion of morality issues. *Social Politics: International Studies in Gender, State & Society, 29*(1), 50–70.

Neto, O. (2023). Semi-presidentialism in Portugal: Academic quarrels amidst institutional stability. In E. J. Fernandes, P. Magalhaes, & A. Pinto (Eds.), *The Oxford handbook of Portuguese politics* (pp. 121–135). Oxford University Press.

Pinto, A., & Paris, A. (2023). Democratization and its legacies. In E. J. Fernandes, P. Magalhaes, & A. Pinto (Eds.), *The Oxford handbook of Portuguese politics* (pp. 18–37). Oxford University Press.

Pratas, M., & Bizzarro, F. (2023). Political parties and party system. In E. J. Fernandes, P. Magalhaes, & A. Pinto (Eds.), *The Oxford handbook of Portuguese politics* (pp. 353–370). Oxford University Press.

Preidel, C., & Knill, C. (2015). Euthanasia: Different moves towards punitive permissiveness. In E. C. Knill, C. Adam, & S. Hurka (Eds.), *On the road to permissiveness?: Change and convergence of moral regulation in Europe* (pp. 79–101). Oxford University Press.

Sabatier, P. (1988). An advocacy coalition framework of policy change and the role of policy-oriented learning therein. *Policy Sciences, 21*, 129–168.

Sabatier, P., & Weible, C. (2007). The advocacy coalition framework innovations and clarifications. In E. P. Sabatier (Ed.), *Theories of the policy process* (2nd ed., pp. 189–220). Routledge.

Salgado, S. (2023). Mass media and political communication. In E. J. Fernandes, P. Magalhaes, & A. Pinto (Eds.), *The Oxford handbook of Portuguese politics* (pp. 308–321). Oxford University Press.

Santos, A. (2018). *A Institucionalização da Bioética e as Políticas Públicas de Saúde em Portugal* (The institutionalization of bioethics and public health policies in Portugal). PhD thesis, Instituto Universitário de Lisboa, Lisboa.

Santos, A. C. (2018). Luta LGBTQ em Portugal: duas décadas de histórias, memórias e resistências [LGBTQ struggle in Portugal: Two decades of stories, memories and resistance]. Transversos: Revista de História, 14, 37–52.

Silva, P. (2020). *Jobs for the Boys? Nomeações para a administração pública.* Fundação Francisco Manuel dos Santos.

Silva, S., Azevedo, L., & Ricou, M. (2019). Determinantes na opinião sobre eutanásia em amostra de médicos portugueses. *Revista Iberoamericana de Bioética, 10,* 1–19.

Stone, D. (2007). Public policy analysis and think tanks. In E. F. Fischer, G. Miller, & M. Sidney (Eds.), *Handbook of public policy analysis: Theory, politics, and methods* (pp. 149–157). CRC Press.

Studlar, D., Burns, G., & Cagossi, A. (2018). Morality policy processes in advanced industrial democracies. *Policy Studies, 39*(5), 479–497.

Studlar, D., Cagossi, A., & Duval, R. (2013). Is morality policy different? Institutional explanations for post-war Western Europe. *Journal of European Public Policy, 20*(3), 353–371.

Tavares, A. (2019). *Administração pública portuguesa.* Fundação Francisco Manuel dos Santos.

Weible, C., & Nohrstedt, D. (2012). The advocacy coalition framework: Coalitions, learning and policy change. In E. E. Araral Jr., S. Fritzen, M. Howlett, M. Ramesh, & X. Wu (Eds.), *Routledge handbook of public policy* (pp. 125–137). Routledge.

CHAPTER 3

The Parliamentary Debate on Euthanasia in Portugal: A Tale in Six Rounds

Abstract This chapter presents the analysis—regarding timing and main involved actors—of six rounds of parliamentary discussion on euthanasia in Portugal: the first which ended in a rejection of four bills; the second, which led to the approval of five bills and a conciliatory text but was stopped by a veto due to a declaration of unconstitutionality by the Constitutional Court; a third, which led to the approval of a revised version of the conciliatory text but was ended by a political veto; a fourth which led to the approval of four bills and a conciliatory text but was stopped by another declaration of unconstitutionality by the Constitutional Court; a fifth that led to the approval of a revised version of the conciliatory text but was also ended by a second political veto; and, a sixth, which included the confirmation of the bill, its promulgation and publication as a Law.

By the end of the chapter, we summarize the main stages of the six rounds and the main contents of the approved Law.

Keywords Euthanasia; Parliament; Petitions; Bills; Political and constitutional vetoes

I. S. Almeida, L. F. Mota, *Politics and Policies in the Debate on Euthanasia*, https://doi.org/10.1007/978-3-031-44588-0_3

This chapter presents the six rounds of parliamentary discussion on euthanasia in Portugal, their timings and the main actors involved in the debates (see timeline at the end of the chapter).

The first round started with the submission of a petition to Parliament by the civic movement 'Right to Die with Dignity' in April 2016 and involved parliamentary discussion of two petitions—this one and an opposing petition—and four bills. The first round ended with a vote and the rejection of those four bills by a small margin in May 2018.

The second round started after the legislative elections of October 2019, with the submission of five bills in late 2019 and early 2020 by five parties. These five bills were then discussed during the course of 2020 and a conciliatory text was approved in January 2021. This joint bill was nevertheless sent to the Constitutional Court by the President and, after a decision pronouncing unconstitutionality due to unclear concepts, the President vetoed the bill in March 2021.

The third round began with the submission of a proposal to change the previous bill by correcting the problems identified by the Constitutional Court. These proposed changes were submitted in early November 2021 and voted on shortly after, in a process that was criticized for being hasty due to the imminent dissolution of Parliament determined by the Government's resignation. The new bill was then rejected by a political veto from the President in the same month.

The fourth round began after the legislative elections of January 2022, with the submission of four bills in early 2022 by the same political parties, as well as a proposal of a referendum by another party. In June 2022, these four bills were approved, while the proposal of the referendum was rejected. After a discussion in the Parliamentary committee, a conciliatory text was approved in December 2022. This document was nevertheless also considered unconstitutional by the Constitutional Court in January 2023.

The fifth round consisted of a proposal which introduced minor changes to the document approved during the fourth round which was jointly presented by the four political parties. This document was approved in March 2023 but was rejected by the President through a political veto.

The sixth round consisted of the confirmation of the bill, obliging the President of the Republic to promulgate it on May 16.

These six rounds will be further discussed in the following subchapters.

3.1   The First Round of the Parliamentary Debate on Euthanasia in Portugal

As stated earlier, the first round of appreciation included the parliamentary discussion of two petitions and four bills, which started in April 2016 and lasted until May 2018.

The first of these petitions, which was in favour of legalizing euthanasia, was submitted to Parliament on April 26, 2016, by the civic movement 'Right to Die with Dignity', created in November 2015. The second petition, titled 'All Lives Have Dignity', was submitted nine months later, on January 25, 2017, by the Portuguese Federation for Life, which, as its title suggests, was against the legalization of euthanasia.

As concerns the bills, they were submitted by four left or centre-left political parties between February 2017 and April 2018: the 'People-Animals-Nature' party (PAN); the Left Bloc (BE); the Socialist Party (PS) and the Green Party (PEV).

The content and the general characteristics of the discussion of these six documents will be analysed in the following sections. After this analysis, a comparison between the four bills is carried out in Sect. 3.1.7, and an analysis of the discussion and voting process is presented in Sect. 3.1.8.

3.1.1   Petition 'Right to Die with Dignity'

Petition No. 103/XIII/1,[1] titled 'Right to Die with Dignity', was submitted to Parliament on April 26, 2016, by the then-renamed 'Civic Movement for the Decriminalization of Assisted Death' with 8427 signatures.

When reading the text of the petition, it can be realized that petitioners advocated the decriminalization and regulation of assisted death, as "*an act to hasten or abbreviate the death of patients in great suffering and without hope of a cure following their own—informed, conscient and reiterated—request*", performed by either assisted suicide or active euthanasia. According to them, the decriminalization of these types of euthanasia is important as a means of exercising the constitutional rights of autonomy, religious freedom and the freedom of belief and conscience, adding that the constitutional right to life should not be considered an irrevocable duty. Likewise, it was also a way to remove religious fundaments from the law in a secular State. Moreover, they considered assisted death to be a compassionate and benevolent act, as it is a way to help people to end unbearable suffering in a dignified way.

According to this document, the petitioners also considered assisted death as a step further in the progressive recognition of patients' rights, such as informed consent, the right to refuse treatments, or the possibility to define their 'living will'. Finally, they mention that the legalization of euthanasia is not in conflict with the promotion of palliative care.

After its submission, this petition was sent to the Parliamentary Committee on Constitutional Affairs, Rights, Freedoms and Guarantees and was admitted on May 4, 2016. The Committee was responsible for assessing it, for which it held 12 public hearings, between June 22, 2016, and July 12, 2016, with the following entities, the content of which will be analysed in Chap. 5: the representative committee of the Civic Movement 'Right to Die with Dignity'; the National Council of Ethics for Life Sciences (CNECV); Jorge Reis Novais and Luísa Neto, two university lecturers in law; the president of the Order of Nurses; the president of the Order of Physicians; the counsellor Judge José Adriano Machado Souto de Moura; Teresa Beleza, a university lecturer in law and, finally, José Francisco de Faria Costa, Mafalda Miranda Barbosa and Manuel Costa Andrade, three university lecturers in law.

Following these hearings, a final report was concluded on November 30, 2016, sent to Parliament and discussed in a parliamentary plenary meeting on February 1, 2017.

3.1.2   Petition 'All Life Has Dignity'

Petition No. 250/XIII/2[2] was submitted to Parliament on January 25, 2017, by the Portuguese Federation for Life, under the title 'All Life has Dignity', with 14,417 signatures. In this petition, its authors begin by considering that *"a Society based on the rule of law and respect for fundamental human rights cannot ignore or silence itself in the face of attempts to threaten the Right to Life, the Dignity, and the concrete Life of each man and woman"*. For these reasons, they argue that human life is the first fundamental right and is thus an inviolable and inalienable right.

The petitioners also mention the State must guarantee and defend the rights of its citizens, while also stressing that *"euthanasia is always a state-supported homicide (allegedly through a health professional) or a state-assisted suicide, and that it is not appropriate to create the right of someone to be killed by others, nor to validate this option as legitimate before the collective".*

They also consider that *"(...) loneliness, vulnerability, and weaknesses are fought with effective social policies, with support and active promotion of hope"* and that *"the associated disease, pain and suffering have remedies that everyone should have access to and that such circumstances do not diminish the Dignity of Human Life, nor take away any value from it".*

Finally, the petitioners take the position that Portuguese policies for fighting the social exclusion of the elderly and disabled, as well as the social answers to terminal patients, are insufficient, while there is also a lack of information or training for citizens on these topics. In conclusion, the petitioners ask Parliament to protect the value of human life—mentioned in article 24 of the Portuguese Constitution—and the Government to reinforce the offer of palliative care in Portugal.

As with the previous petition, this second petition was sent to the Committee on Constitutional Affairs, Rights, Freedoms, and Guarantees and admitted on February 1, 2017. The discussion of this petition involved eighteen public hearings held between April 19, 2017, and February 9, 2018, with the following entities: the Portuguese Federation for Life; the National Council of Ethics for Life Sciences (CNECV); the Order of Lawyers; António Cluny, deputy attorney general; the civic movement 'STOP Euthanasia'; the Director General of Health; the Portuguese Association of Palliative Care; the National Commission for Justice and Peace; the Civic Movement for the Decriminalization of Assisted Death; Guilherme da Fonseca, retired counselling judge; the university lecturers in law Cristina Líbano Monteiro, Inês Fernandes Godinho, Inês Ferreira Leite and Tiago Duarte and the physicians João Oliveira (MD), José Manuel de Paiva Jara (psychiatrist) and Ramon de La Féria (surgeon). A final report was concluded on May 24, 2018, a few days before the bills were voted and sent to Parliament on July 10, after the vote on the bills.

3.1.3 The Bill from the People-Animals-Nature (PAN) Party

The first bill trying to regulate euthanasia was submitted to Parliament by **PAN** on February 21, 2017—Bill No. 418/XIII/2ᵃ.[3] After its submission, the bill was sent to the Committee on Constitutional Affairs, Rights, Freedoms, and Guarantees (in conjunction with the Health Committee). Both committees issued opinions on March 22, 2017. Before these committee opinions, on March 9, 2017, written opinions were requested from the following entities: the Order of Physicians, the Order of Lawyers, the High Council of the Public Prosecution Service, the Order of Nurses, the High Council of the Judiciary and the National Council of Ethics for Life Sciences (which replied), as well as the Order of Psychologists and the College of the Specialty of Psychiatry of the Order of Physicians, which did not reply.

3.1.4   The Bill from the Left Bloc (BE)

The **BE** presented its bill on February 7, 2018—Bill No. 773/XIII/3ª.[4] After its submission, this bill was sent to the Committee on Constitutional Affairs, Rights, Freedoms and Guarantees (in conjuction with the Health Committee). The Committee issued an opinion on May 22, 2018, after expert opinions were requested on February 15, 2018, from the following entities: the High Council of the Judiciary, the Order of Lawyers, the Order of Portuguese Psychologists and the Order of Nurses (which replied), as well as the High Council of the Public Prosecution Service; the Order of Physicians and the National Ethics Council for Life Sciences, which did not reply.

3.1.5   The Bill from the Socialist Party (PS)

The **PS** presented Bill No. 832/XIII/3ª[5] on April 13, 2018. Like the previously mentioned bills, the bill submitted by PS was sent to the same committees. The committees issued an opinion on May 23, 2018, after opinions were requested on April 19, 2018, from the following entities: the Order of Nurses, the Order of Lawyers, the Order of Portuguese Psychologists (which replied), as well as the High Council of the Public Prosecution Service, the High Council of the Judiciary, the Order of Physicians and the National Council of Ethics for Life Sciences, which did not reply.

3.1.6   The Bill from the Green Party (PEV)

The fourth bill was Bill No. 838/XIII/3ª,[6] submitted on April 20, 2018, by the **PEV**. After its submission, the bill was sent to the same committees, which issued their opinion on May 23, 2018, after opinions were requested on April 26, 2018, from the following entities: the Order of

Lawyers and the Order of Portuguese Psychologists (which replied), as well as to the High Council of the Judiciary, the High Council of the Public Prosecution Service, the Order of Nurses, the Order of Physicians and the National Council of Ethics for Life Sciences, which did not reply.

3.1.7 Comparison of the Four Bills

The four submitted bills do not differ substantively regarding the definition of assisted death or the conditions to make the request, as can be seen in Fig. 3.1.

Bill 418/XIII/2ª (PAN)	Bill 773/XIII/3ª (BE)	Bill 832/XIII/3ª (PS)	Bill 838/XIII/3ª (PEV)
Main similarities			
• Only for adult people; • Health professionals may be conscientious objectors; • Targeting cases of extreme, unbearable suffering, with no hope of healing and incurable and fatal disease; • The requests must be current, free, conscious, reiterated, and enlightened; • The request may only be made by patients themselves to a physician.			
The main differences			
The evaluation committee is composed of physicians, jurists, and a personality in ethics or bioethics.	The evaluation committee is composed of jurists, health professionals and three specialists in ethics or bioethics.	The evaluation committee is composed of jurists, physicians, nurses, and a bioethics specialist appointed by the CNECV.	The evaluation committee is regionally oriented and is composed of physicians, nurses, lawyers, and a judge.
The patient may revoke the process at any time and the process is stopped if the patient becomes unconscious.	The patient may revoke the process at any time and the process is stopped if the patient becomes unconscious unless the Living Will specifies differently.	The patient may revoke the process at any time and the process is stopped if the patient becomes unconscious.	The patient may revoke the process at any time and the process is stopped if the patient becomes unconscious.
Performed in all hospitals.	Performed in all hospitals.	Performed in all hospitals.	Performed only in public hospitals.

Fig. 3.1 Main differences and similarities of the four bills from the 'first round'. (Source: own elaboration, based on the information provided on the Parliament's website [The full comparison of bills developed by parliamentary services is available at: https://www.parlamento.pt/Documents/2018/Maio/mapacomparativo.pdf])

For instance, the four bills state that only Portuguese citizens or legal residents who are older than 18 years can submit a request for assisted death. Moreover, they all specify the request must be made by patients themselves following a conscious, free and enlightened decision and confirmed several times throughout the process. In addition, all bills accept that health professionals should not be obliged to participate in assisted dying, thus guaranteeing their right to conscientious objection.

Likewise, all of them define approximately the same following process:

1. The patient issues a written request to a physician of their choice (assisting physician).
2. This physician assesses the request, checks if all the criteria are met and informs patients about their health conditions and the existing therapeutic alternatives.
3. A physician with expertise in the patient's pathology is consulted.
4. The patient is examined by a psychiatrist if there are any doubts about their capacity to decide or their mental health.
5. An Evaluation Committee is consulted to make sure all the criteria are met.
6. The patient, together with the assisting physician, chooses the method of death (assisted suicide or active euthanasia) and the date and place of death (health facilities or the patient's home), as well as who can be present during the act of death.

All bills also specified that the process should be revoked if any intervening physicians have a negative opinion or if patients do not reiterate their decision throughout the process. Likewise, all bills specified that the process is stopped if patients become unconscious.

The main difference (see Fig. 3.1) between the bills concerns the place where assisted death would be performed, since the PEV considered that only public hospitals should be able to carry out euthanasia procedures. As regards the revocation of the process, the bill from the BE, unlike the others, specified that the process may not be revoked in case the patient becomes unconscious if that situation is specified in the Living Will.

Another difference is the composition of the 'Evaluation Committee'. The PAN proposed the creation of a 'Law Enforcement Control and Evaluation Committee', composed of physicians, lawyers and a professional in the ethical or bioethical area. The BE proposed a committee composed of jurists, health professionals and three specialists in ethics or bioethics. The committee proposed by the PS is composed of physicians, nurses, jurists and bioethics professionals appointed by the CNECV. Finally, the PEV proposed committees for each Regional Health Administration (NUT II), evaluating the processes by region, composed of physicians, nurses, lawyers and a judge.

3.1.8   Discussion and Voting of the Bills of the 'First Round'

After their discussion in the Committee on Constitutional Affairs, Rights, Freedoms and Guarantees, these four bills were discussed in a parliamentary plenary session held on May 29, 2018. During this debate, the four bills were presented and MPs from all the political parties intervened, in what can be named as a 'heated debate', which will be further analysed in the next chapter.

After the debate, the bills were voted on and rejected by a small margin, as can be seen in the following Fig. 3.2.

As can be seen, the bill which was closest to being approved was that from the PS, but the results do not differ significantly. This bill had votes in favour of MPs from the PS (except two members), the BE, the PEV and PAN, and votes against from MPs from the PSD (except four members), the CDS-PP and the PCP.

It should be noted that, during the voting process, some MPs stated that they would submit an 'explanation of vote' to explain their voting positions. This is particularly important when there is a voting discipline imposed by the parliamentary groups of the political parties. In this specific vote, only the PSD and the PS gave the freedom to vote to their MPs.

	Bill 418/XIII/2ᵃ (PAN)	Bill 773/XIII/3ᵃ (BE)	Bill 832/XIII/3ᵃ (PS)	Bill 838/XIII/3ᵃ (PEV)
In favour	102	104	110	104
Against	116	117	115	117
Abstentions	11	8	4	8
Total	229	229	229	229

Fig. 3.2   Results of the vote of the four bills from the 'first round'. (Source: Parliament website [All details regarding the voting on all bills are available at: https://www.parlamento.pt/Paginas/2018/maio/MorteAssistida.aspx])

After this result, the topic would be resumed only after the legislative elections of October 2019, in a 'second round' that will be described in the following section.

3.2   THE SECOND ROUND OF THE PARLIAMENTARY DEBATE ON EUTHANASIA IN PORTUGAL

As mentioned earlier, the second round of discussion included the parliamentary discussion of five bills, starting in October 2019, after the legislative elections of October 6, 2019, and ending in April 2021, when the President vetoed the bill that was approved by Parliament.

The first bills were submitted by the same four parties which had submitted bills in the previous legislature between October and December 2019: the Left Bloc (BE); the People-Animals-Nature party (PAN); the Socialist Party (PS) and the Green Party (PEV). The fifth bill was submitted in February 2020 by the party Liberal Initiative (IL). The content and the general characteristics of the discussion of these five documents will be analysed in the following sections. After that, a comparison of them will be carried out in Sect. 3.2.6, and an analysis of the discussion and voting process will be presented in Sect. 3.2.7.

3.2.1   The Bill from the Left Bloc (BE)

The first bill from the 'second round' was submitted on October 25, 2019—Bill No. 4/XIV/1.[7] After its submission, the bill was sent to the same committee, which issued its opinion on February 12, 2020, but, before that, written opinions were requested from the following entities on November 15, 2019: the Order of Nurses, the Order of Physicians, the High Council of the Judiciary, the High Council of the Public Prosecution Service and the National Council of Ethics for Life Sciences (which all replied), as well as the Order of Lawyers, which replied only with a short piece of information.

3.2.2   The Bill from the People-Animals-Nature Party (PAN)

PAN presented its bill on November 12, 2019—Bill No. 67/XIV/1.[8] After its submission, the bill was also sent to the committee, which issued an opinion on February 20, 2020, after opinions were requested from the following entities on November 20, 2019: the Order of Nurses, the High Council of the Public Prosecution Service, the Order of Physicians, the Order of Lawyers, the High Council of the Judiciary and National Council of Ethics for Life Sciences (which all replied), as well as the Order of Lawyers, which replied only with a short piece of information.

3.2.3   *The Bill from the Socialist Party (PS)*

The Socialists submitted Bill No. 104/XIV/1[9] on November 21, 2019. This bill was also sent to the committee, which delivered its opinion on February 20, 2020, after opinions were requested on November 27, 2019, from the following entities: the Order of Psychologists, the Order of Nurses, the Order of Physicians, the High Council of the Judiciary, the High Council of the Public Prosecution Service and the National Council for Ethics in Life Sciences (which all replied).

3.2.4   *The Bill from the Green Party (PEV)*

The Green Party submitted a new bill on December 13, 2019, designated Bill No. 168/XIV/1,[10] which followed the same ideas as the one the party presented in 2018. The bill was also sent to the same committee, which issued an opinion on February 12, 2020, after opinions were requested on December 23, 2019, from the following entities: the Order of Psychologists, the Order of Nurses, the High Council of the Judiciary, the Order of Physicians, the National Council for Ethics in Life Sciences and the High Council of the Public Prosecution Service (which all replied).

3.2.5   *The Bill from the Liberal Initiative (IL)*

The last bill was presented on February 3, 2020, by the IL—Bill No. 195/XIV/1.[11] This bill was also sent to the same committee, and its analysis included a request for written opinions from the following entities on February 12, 2020: the Order of Portuguese Psychologists, the Order of Nurses, the Order of Lawyers, the Order of Physicians and the High Council of the Public Prosecution Service (which all replied), as well as the High Council of the Judiciary, which issued only a short piece of information.

3.2.6   Comparison of Bills

As in the 'first round', the five bills were significantly similar, specially concerning the conditions to require the process and the fact that health professionals could be conscious objectors—see Fig. 3.3.

Bill 4/XIV/1 (BE)	Bill 67/XIV/1 (PAN)	Bill 104/XIV/1 (PS)	Bill 168/XIV/1 (PEV)	Bill 195/XIV/1 (IL)
Main similarities				
• Only for adult people; • Health professionals may be conscientious objectors; • Targeting cases of extreme and unbearable suffering, with no hope of healing and incurable and fatal disease; • The requests must be current, free, conscious, reiterated, and enlightened; • The request may only be made by patients themselves to a physician.				
Main differences				
The evaluation committee is composed of jurists, health professionals and three specialists in ethics or bioethics.	The evaluation committee is composed of physicians, jurists, and a personality in ethics or bioethics.	The evaluation committee is composed of jurists, physicians, nurses, and a bioethics specialist appointed by the CNECV.	The evaluation committee is composed of physicians, nurses, lawyers, and a magistrate.	The evaluation committee is composed of jurists, physicians, nurses, and a bioethics specialist appointed by the CNECV.
The patient may revoke the process at any time and the process is stopped if the patient becomes unconscious unless the Living Will specifies differently	The patient may revoke the process at any time and the process is stopped if the patient becomes unconscious	The patient may revoke the process at any time and the process is stopped if the patient becomes unconscious	The patient may revoke the process at any time and the process is stopped if the patient becomes unconscious.	The patient may revoke the process at any time and the process is stopped if the patient becomes unconscious.
Performed in all hospitals	Performed in all hospitals	Performed in all hospitals.	Performed only in public hospitals	Performed in all hospitals

Fig. 3.3 Main differences and similarities of the four bills from the 'second round'. (Source: own elaboration, based on the content of the bills)

Likewise, the main differences remain the same. The composition of the evaluation committee is slightly different between the bills, the bill from the PEV only enables assisted death to be performed in public hospitals, and the bill from the BE specifies less strict conditions for revocation.

3.2.7   Discussion and Voting of the Bills and Conciliatory Text

After the discussion of the five bills in the committee, they were discussed at a plenary meeting held on February 20, 2020, again in a 'heated debate'. After the debate, the five bills were voted on and approved by a small margin—see Fig. 3.4.

The bill from the PS, which was approved with the highest margin, had favourable votes from the MPs of the PS (except for eight members), the BE, the PEV, PAN, and the IL and votes against from the members of the PSD (except for eleven members), the CDS, the PCP and CH. Once again, several MPs submitted 'explanations of vote' to explain their votes.

After the generic approval of these five bills, they were sent to the committee to create a conciliatory text. This process—which was delayed by the beginning of the pandemic situation—included six hearings, which were held between July 2020 and January 2021, with: the Association of Portuguese Catholic Physicians; the Interreligious Working Group on Religion & Health; the association Together for Life; the Portuguese Federation for Life; the Voiceless Children Movement and the STOP Euthanasia movement.

The final conciliatory text—Decree 109/XIV[12]—which largely followed the criteria and process specified in the bill proposed by the PS, thus specified that one should consider *"a non-punishable anticipation of assisted death that which occurs by decision of the adult person whose will is*

	Bill 4/XIV/1 (BE)	Bill 67/XIV/1 (PAN)	Bill 104/XIV/1 (PS)	Bill 168/XIV/1 (PEV)	Bill 195/XIV/1 (IL)
In favour	124	121	128	114	114
Against	84	85	84	85	84
Abstentions	14	16	10	23	24
Total	222	222	222	222	222

Fig. 3.4   Results of the vote of the four bills from the 'second round'. (Source: Own production)

current and reiterated, serious, free and enlightened, in a situation of intolerable suffering, with a definitive injury of extreme gravity according to the scientific consensus or incurable and fatal illness, when practised or helped by health professionals" (art. 2, nr. 1).

After confirmatory voting, this text was then sent to the President on February 10, 2021.

3.2.8   Rejection by the Constitutional Court

After a first reading, the President decided to send the decree to the Constitutional Court, arguing in his statement[13] that he was not asking the Court to assess if the concept of euthanasia was in line with the Constitution, but rather if the proposed regulation of euthanasia was aligned with it. Later in the document, the President expressed particular concern about some concepts being "excessively indeterminate", giving the example of the concepts of the "situation of intolerable suffering" and "definitive injury of extreme gravity according to scientific consensus".

In a statement[14] issued on March 15, 2021, the Court declared the decree unconstitutional as it considered the concept of definitive injury of extreme gravity (unlike that of intolerable suffering) was indeed undetermined but claimed that euthanasia, as a whole, was not against the Constitution.

Following this decision taken by the Constitutional Court, the President resent the Decree to Parliament without promulgation.[15] Considering this outcome, Parliament could correct the bill and resubmit it to the President, which was what happened in the 'third round' of discussion described in the following section.

3.3   THE THIRD ROUND OF THE PARLIAMENTARY DEBATE ON EUTHANASIA IN PORTUGAL

The third round of debate started in November 2021, with the discussion of a new bill which made amendments to the previously vetoed bill and ended with a presidential veto of the resulting bill.

3.3.1   *The Approval of a Revised Version*

Following the declaration of unconstitutionality of the decree approved in the second round, the five political parties (BE, PEV, PS, PAN and IL) came together to create an amending bill—which, after its approval, became Decree 199/XIV[16]. In this, the legislators decided to use the expression 'medically assisted death' rather than 'anticipation of death' and to present a list of definitions in article 2, in which the following, strongly connected with the Constitutional Court's declaration of unconstitutionality, can be highlighted:

> *"d) Serious or incurable illness: a serious illness that threatens life, in an advanced phase, that is also progressive, incurable and irreversible, which causes suffering of great intensity;*
>
> *e) Definitive injury of extreme severity: a serious, definitive and largely disabling injury that places the person in a situation of dependence on a third party or technological support to carry out elementary activities of daily living, with certainty or very high probability that such limitations will come to persist over time with no possibility of cure or significant improvement;*
>
> *f) Suffering: physical, psychological and spiritual suffering, resulting from a serious or incurable disease or from a definitive injury of extreme gravity and of great intensity, and that is persistent, continuous or permanent and is considered intolerable by the person"*

After these general concepts, the bill thus defined non-punishable medically assisted death as "that which occurs *by decision of the people themselves, adult, whose will is current and reiterated, serious, free and enlightened, in a situation of intolerable suffering, with a definitive injury of extreme gravity or incurable and fatal illness, when practised or helped by health professionals"* (art. 3, nr. 1).

This bill was once again discussed in a plenary meeting and a vote was held on November 5, 2021, with the results presented in Fig. 3.5.

Fig. 3.5 Results of the vote of the conciliatory bill from the 'third round'. (Source: Own production)

Bill 199/XIV (conciliatory text)	
In favour	138
Against	84
Abstentions	5
Total	227

The votes in favour were from the PS (except for seven members), the BE, the PEV, PAN, the IL and two independent MPs, and the votes against from the PSD (except for thirteen members), the CDS, the PCP and CH. Once again, some MPs provided an 'explanation of vote', as only the PS and the PSD gave freedom of vote to their members.

On November 26, 2021, this bill was sent to the President.

3.3.2   *The Presidential Veto*

Shortly after receiving the decree, the President decided to veto[17] it on November 30, 2021. The reasons given by the President for this political veto—since the Constitutional Court was not involved—were the fact that, in article 3 nr 1, the bill specified that one of the conditions to a valid euthanasia request was the patient having an "incurable and fatal illness", while in article 3 nr 3 (and in article 2, paragraph d) specified that one of the conditions is having a "serious or incurable illness". Therefore, the President mentioned Parliament needed to clarify this, as the new version was much more radical than the previous one. He also felt the need to stress that this decision was not related to *"(…) any personal religious, ethical, moral, philosophical or political position (…)"*, as he is known for being a devoted Catholic and from a right-wing political party.

This decision could have led to a new clarification from Parliament but the dissolution of Parliament, following the Government's resignation in late October 2021, prevented that possibility. Therefore, the discussion of this issue had to be resumed after the new legislative elections and with a process which had to start over once again, in a fourth round that will be briefly described in the following section.

3.4   THE FOURTH ROUND OF THE PARLIAMENTARY DEBATE ON EUTHANASIA IN PORTUGAL

The fourth round of discussion of euthanasia started shortly after the legislative elections of January 30, 2022, with the submission of four new bills on the following dates: from the BE on March 29, 2022—Bill 5/XV/1[18]; from the PS on May 17, 2022—Bill 74/XV/1[19]; from PAN on May 20, 2022—Bill 83/XV/1[20] and from the IL on June 2, 2022—Bill 111/XV/1.[21] In this regard, it should be stressed that the PEV did not submit a bill again as it did not have any seats in the 'newly formed' parliamentary composition. Besides these four bills which aimed to decriminalize euthanasia, the Enough! party (CH) decided to propose the organization of a referendum on the topic of euthanasia.

As revealed in the 'raison d'être' of each of these bills, they follow the ideas that had been expressed in the conciliatory texts from the second and third rounds of discussion. It should be stressed nonetheless that the bills from the BE, PS and IL opted to use the expression "serious and incurable illness", while the bill from PAN used the expression "serious or incurable illness". The main differences were those presented in the second-round bills concerning the composition of the evaluation committee (see point 3.2.6).

All these bills were sent to the Committee on Constitutional Affairs, Rights, Freedoms and Guarantees. They were then discussed with the help of requested written opinions issued by the following entities: the Order of Lawyers; the High Council of the Judiciary; the Order of Physicians; the Order of Nurses; the High Council of the Public Prosecution Service; the National Council of Ethics for Life Sciences and the Order of Psychologists.

3.4.1   The Proposal of the Referendum from the Enough! Party (CH)

On May 20, 2022, the far-right Enough! party submitted Resolution Project No. 62/XV/1,[22] which aimed to promote the organization of a referendum about euthanasia.

According to this document, *"the Constitution of the Portuguese Republic establishes (…), that human life and the moral and physical integrity of people are inviolable [adding that] the guarantee of the right to life is, in fact, the presupposition and condition par excellence for the realization of all other fundamental rights".* Under these circumstances and given the fact that most European countries still criminalize euthanasia, the proponents of this document were against the previously mentioned bills, stating that the party *"does not accept nor tolerates the idea that there are lives which are worth living and others are not".*

Therefore, this document proposed that a referendum should be organized, in which citizens should be asked to answer the following question: *"Do you agree that the medically assisted death of a person, at their request, or an assisted suicide, should continue to be punished by criminal law?".*

	Bill 5/XV/1.ª (BE)	Bill 24/XV/1.ª (PS)	Bill 83/XV/1.ª (PAN)	Bill 111/XV/1.ª (IL)	Resolution 62/XV/1.ª (CH)
In favour	127	128	126	127	71
Against	88	88	88	88	147
Abstentions	6	5	7	6	2
Total	221	221	221	221	221

Fig. 3.6   Results of the vote on the four bills and the resolution from the 'fourth round'. (Source: Own production)

3.4.2   Voting and Discussion of the Bills of the 'Fourth Round'

The four bills were then officially presented in a parliamentary plenary meeting held on June 9, 2022. These four bills were all approved with very similar results (see Fig. 3.6)—the one from the PS received votes in favour from MPs of the PS (except for seven members), the BE, IL, PAN and the Free party (L) and votes against from the members of the PSD (except for six members), the PCP and CH. On the other hand, the Resolution Project from CH was rejected by a very large margin. Once again, it is important to mention that only two parties—the PS and the PSD—gave their members freedom of vote, which explains why several MPs presented an 'explanation of vote'.

After the approval of these four bills, they went back to the committee to be further discussed and so that a conciliatory text could be created. During this discussion, several entities decided to issue a written statement or to ask for a hearing, namely the following: the Portuguese Union Seventh Day Adventists; the Portuguese Buddhist Union; the Lisbon Jewish Community; the DNA party; the Centre of Bioethics Studies; the Portuguese Association of Catholic Psychologists; the VivaHáVida (HurrayForLife) Association; the InFamily association; Association for the Defence and Support of Life—Aveiro; the citizen João Emanuel Diogo; the lawyer Teresa de Melo Ribeiro and the Portuguese Association of Insurers. The in-person hearings were held between September and October, with the following entities: the Portuguese Association of Catholic Psychologists; the HurrayForLife Association; InFamily; the Association for the Defence and Support of Life—Aveiro and the Speciality Council in Clinical and Health Psychology from the Order of Portuguese Psychologists.

After this debate, a conciliatory text was then approved, which largely followed the criteria and process specified in the bill proposed by the PS, which changed the following concepts in comparison to the previous decree (changes or additions are underlined):

> *d) "Serious __and__ incurable illness": An incurable and irreversible illness that threatens life in an advanced and progressive phase, and which causes suffering of great intensity.*
>
> *f) "Suffering of __Great Intensity__": physical, psychological and spiritual suffering, resulting from a serious __and__ incurable disease or a definitive injury of extreme gravity, with great intensity, and that is persistent, continuous or permanent and it is considered intolerable by the person themselves.*

Under these circumstances, this new bill considered non-punishable medically assisted death *"(…) that which occurs by decision of the adult people themselves, whose will is current and reiterated, serious, free and enlightened, in a situation of intolerable suffering, with a definitive injury of extreme gravity or __serious and incurable__ illness, when practised or helped by health professionals"* (art. 3, nr. 1).

The conciliatory text was then voted on and approved on December 9, 2022, despite a request for postponement by CH. It should be noted that only four days before this final vote, the PSD presented Draft Resolution 311/XV/1[23] also expressing the intention of holding a referendum. This draft resolution was nevertheless rejected on December 7 by the President of Parliament, who based his decision on the fact that a legislative initiative with the same goal had already been presented—the draft resolution of CH—and rejected.

The conciliatory text—Decree Nr. 23/XV[24]—was then sent to the President on January 4, 2023.

3.4.3 The (Second) Rejection by the Constitutional Court

After receiving the new decree, the President decided to send it to the Constitutional Court, arguing in its statement[25] that this court should analyse whether this new decree has solved the issues of unclear concepts it had raised in its 2021 appreciation.

In a statement[26] issued on January 30, 2023 (Judgement 5/2023), the court vetoed the decree, stating that Parliament had significantly altered the meaning of the law, arguing that the text generates uncertainty as to whether the degree of suffering to resort to medically assisted death is 'cumulative'—physical, psychological and spiritual—or whether just one of these dimensions is required.

Following this decision taken by the Constitutional Court, the President resent the decree to Parliament without promulgation.[27] Considering this outcome, Parliament had the possibility of correcting the bill and resubmitting it to the President, in a process we will describe in the following subsection.

3.5   THE FIFTH ROUND OF THE PARLIAMENTARY DEBATE ON EUTHANASIA IN PORTUGAL

The fifth round of discussion of euthanasia consisted of a proposal of slight changes to the previous version, which was approved on March 31, 2023, but was again vetoed by the President on April 19, 2023.

3.5.1   *The Approval of a Revised Version*

Taking into consideration the declaration of unconstitutionality of the decree which had been approved in the fourth round, the four political parties which had authored the decree (BE, PS, PAN and IL) created an amending bill, which later became Decree 43/XV.[28] This new bill contained a new definition of "suffering of great intensity", since the Constitutional Court asked legislator to clarify whether there was a need for patients to have accumulated "physical, psychological and spiritual suffering" or just some of them. Therefore, the authors of this new bill decided to simplify the concept and define it as such:

> *"Suffering of <u>Great Intensity</u>": the suffering resulting from a serious incurable disease or from a definitive injury of extreme gravity, with great intensity, that is persistent, continuous or permanent and is considered intolerable by the person themselves.*

In addiction to this slight change, this amending bill also introduced a significant change, by suggesting that medically assisted death could only occur through euthanasia if assisted suicide was impossible due to the patient's physical disability, thus adding article 3(5), which specifically stated that:

> *Medically assisted death can only occur by euthanasia when medically assisted suicide is impossible due to the patient's physical disability.*

Fig. 3.7 Results of the vote of the four bills and the resolution from the 'fifth round'. (Source: Own production)

	Decree 43/XV (1ˢᵗ time)
In favour	135
Against	87
Abstentions	2
Total	224

The bill was then officially presented in a parliamentary plenary meeting held on March 31, 2023. After the debate, the bill was approved with very similar results (see Fig. 3.7): the votes were in favour from MPs of the PS (except for six members), the BE, IL, PAN and the Free party (L) and votes against from the members of the PSD (except for seven members), the PCP and CH.

After the approval of the decree, it was published on March 13, 2023, and was again sent to the President of the Republic.

3.5.2   *The (Second) Presidential Veto*

After receiving the decree, the President decided to veto it[29] and asked *"the Assembly of the Republic to consider clarifying who defines the patient's physical incapacity to self-administer lethal drugs, as well as who should ensure medical supervision during the act of medically assisted death..."*

Considering this veto on April 19, 2023, the decree was resent to Parliament, for it to be amended once again.

3.6   The Sixth Round of the Parliamentary Debate on Euthanasia in Portugal

The sixth round of discussion on euthanasia consisted of the new review of the decree and its confirmation, on May 12, 2023, obliging the President of the Republic to promulgate it on May 16.

3.6.1   *The Confirmation of the Bill*

Taking into consideration the veto, the four political parties which had authored the decree (BE, PS, PAN and IL) decided to confirm the existing bill and introduce no changes in the decree.

Fig. 3.8 Results of the vote of the four bills and the resolution from the 'sixth round'. (Source: Own production)

	Decree 43/XV (2nd time)
In favour	129
Against	81
Abstentions	1
Total	211

The bill was then officially presented in a parliamentary plenary meeting held on May 12, 2023. After the debate, the bill was approved with very similar results (see Fig. 3.8): the votes were in favour from MPs of the PS (except for four members), the BE, IL, PAN and the Free party (L) and votes against from the members of the PSD (except for eight members), the PCP and CH.

After the approval of the decree, it was published on May 12, 2023, and was again sent to the President of the Republic. The President was obliged to promulgate, which happened on May 16.

3.6.2 Promulgation

After receiving the decree, the President was obliged to promulgate the decree, following article 136, paragraph 2 of the Constitution, since it imposes that, after the veto of a decree, if Parliament "(…) confirms its vote by an absolute majority of all the Members of the Assembly of the Republic in full exercise of their office, the President of the Republic must enact the legislative act within a time limit of eight days counting from its receipt".

After the promulgation from the President on May 16, 2023, the decree was sent to Parliament for final drafting and was published in the official gazette (Diário da República) on May 25, 2023.

3.7   THE FINAL VERSION OF THE LAW

As can be seen throughout the chapter, the parliamentary debate on euthanasia was particularly heated with six rounds of discussions, which are summarized in the Fig. 3.9.

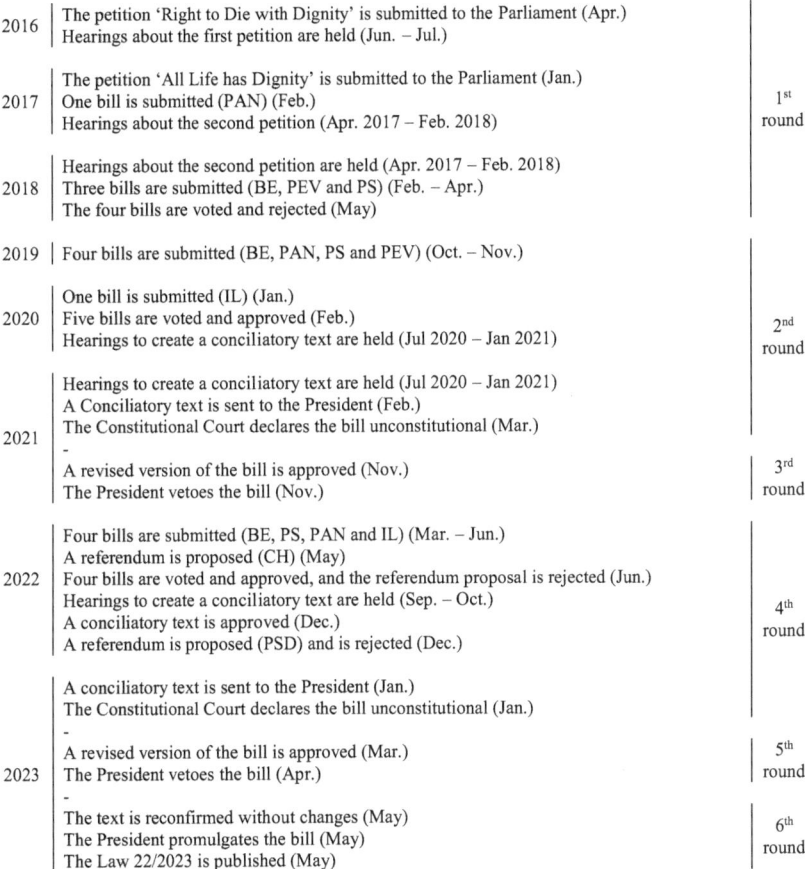

2016	The petition 'Right to Die with Dignity' is submitted to the Parliament (Apr.) Hearings about the first petition are held (Jun. – Jul.)	
2017	The petition 'All Life has Dignity' is submitted to the Parliament (Jan.) One bill is submitted (PAN) (Feb.) Hearings about the second petition (Apr. 2017 – Feb. 2018)	1st round
2018	Hearings about the second petition are held (Apr. 2017 – Feb. 2018) Three bills are submitted (BE, PEV and PS) (Feb. – Apr.) The four bills are voted and rejected (May)	
2019	Four bills are submitted (BE, PAN, PS and PEV) (Oct. – Nov.)	
2020	One bill is submitted (IL) (Jan.) Five bills are voted and approved (Feb.) Hearings to create a conciliatory text are held (Jul 2020 – Jan 2021)	2nd round
2021	Hearings to create a conciliatory text are held (Jul 2020 – Jan 2021) A Conciliatory text is sent to the President (Feb.) The Constitutional Court declares the bill unconstitutional (Mar.) - A revised version of the bill is approved (Nov.) The President vetoes the bill (Nov.)	3rd round
2022	Four bills are submitted (BE, PS, PAN and IL) (Mar. – Jun.) A referendum is proposed (CH) (May) Four bills are voted and approved, and the referendum proposal is rejected (Jun.) Hearings to create a conciliatory text are held (Sep. – Oct.) A conciliatory text is approved (Dec.) A referendum is proposed (PSD) and is rejected (Dec.)	4th round
2023	A conciliatory text is sent to the President (Jan.) The Constitutional Court declares the bill unconstitutional (Jan.) - A revised version of the bill is approved (Mar.) The President vetoes the bill (Apr.) - The text is reconfirmed without changes (May) The President promulgates the bill (May) The Law 22/2023 is published (May)	5th round 6th round

Fig. 3.9 Timeline of the discussion of euthanasia. (Source: Own production)

After these six rounds of discussion, the resulting law thus specified that people may ask for medically assisted death under the condition that:

   a. they are adults;
   b. they are national citizens or legal residents in the national territory.
   c. their will to ask for it is current and reiterated, serious, free, and clear, and expressed through a decision that is personal, non-delegable and revocable at any time;

d. they are in a situation of suffering of great intensity, that is "suffering resulting from a serious and incurable illness or permanent injury of extreme gravity, with great intensity, persistent, continuous or permanent and considered intolerable by the person";

e. they have a definitive injury of extreme severity or a serious and incurable illness, that is a "disease that threatens life, in an advanced and progressive phase, incurable and irreversible, which causes suffering of great intensity" or a "serious, definitive and largely disabling injury that places the person in a situation of dependence on a third party or technological support to carry out the elementary activities of daily life, with very high certainty or probability that such limitations will persist over time without the possibility of cure or significant improvement";

As concerns the process, the law specified the following steps:

1. Patients issue a written request to a physician of their choice (guiding physician) (article 4);
2. The guiding physician assesses the request, checks if all the criteria are met and informs patients about their health conditions and the existing therapeutic alternatives (article 5);
3. A physician with expertise in the patient's pathology is consulted (article 6);
4. Patients are examined by a psychiatrist if there are any doubts about their capacity to decide or their mental health (article 7);
5. A Verification and Assessment Committee is consulted to make sure all the criteria are met (article 8);
6. The patient, together with the guiding physician, chooses the method of death (assisted suicide or active euthanasia), the date and place of death (health facilities or the patient's home), as well as who can be present during the act of death besides the guiding physician and other healthcare professional (articles 9 and 10);
7. The administration of the lethal drug (article 10).

Apart from those criteria and this process, the law also specifies that medically assisted death should be preferably carried out as assisted suicide (article 3), which may be done in public or private hospitals (article 13), that the involved health professionals should maintain professional secrecy and confidentiality (article 14) and that all health professionals have the right to be conscientious objectors (article 21). Finally, it is important to stress that the entire process shall be monitored by the General Inspectorate of Healthcare Activities (IGAS) and a Committee of Verification and Assessment of Clinical Procedures of Medically Assisted Death (CVA), composed of the following members: (a) a lawyer appointed by the High Council for the Judiciary; (b) a jurist appointed by the High Council of the Public Prosecution Service; (c) a physician appointed by the Order of Physicians; (d) a nurse designated by the Order of Nurses; and (e) a specialist in bioethics designated by the National Council of Ethics for Life Sciences (article 25).

NOTES

1. All details regarding this petition are available at https://www.parlamento.pt/ActividadeParlamentar/Paginas/DetalhePeticao.aspx?BID=12783
2. All details regarding this petition are available at: https://www.parlamento.pt/ActividadeParlamentar/Paginas/DetalhePeticao.aspx?BID=12931
3. All details regarding this bill are available at https://www.parlamento.pt/ActividadeParlamentar/Paginas/DetalheIniciativa.aspx?BID=41038
4. All details regarding this bill are available at https://www.parlamento.pt/ActividadeParlamentar/Paginas/DetalheIniciativa.aspx?BID=42165
5. All details regarding this bill are available at https://www.parlamento.pt/ActividadeParlamentar/Paginas/DetalheIniciativa.aspx?BID=42453
6. All details regarding this bill are available at: https://www.parlamento.pt/ActividadeParlamentar/Paginas/DetalheIniciativa.aspx?BID=42476
7. All details relating to this bill are available at: https://www.parlamento.pt/ActividadeParlamentar/Paginas/DetalheIniciativa.aspx?BID=43947
8. All details regarding this bill are available at: https://www.parlamento.pt/ActividadeParlamentar/Paginas/DetalheIniciativa.aspx?BID=44066
9. All details regarding this bill are available at: https://www.parlamento.pt/ActividadeParlamentar/Paginas/DetalheIniciativa.aspx?BID=44151
10. All details regarding this bill are available at: https://www.parlamento.pt/ActividadeParlamentar/Paginas/DetalheIniciativa.aspx?BID=44311
11. All details regarding this bill are available at: https://www.parlamento.pt/ActividadeParlamentar/Paginas/DetalheIniciativa.aspx?BID=44422
12. All details regarding this conciliatory bill are available at https://www.parlamento.pt/ActividadeParlamentar/Paginas/DetalheDiplomaAprovado.aspx?BID=22631
13. All details regarding this statement from the President are available at: https://www.presidencia.pt/atualidade/toda-a-atualidade/2021/02/

presidente-da-republica-submete-decreto-da-eutanasia-ao-tribunal-constitucional/

14. All details regarding this decision from the Constitutional Court are available at: https://www.tribunalconstitucional.pt/tc/imprensa0200-bd5860.html

15. All details regarding this statement from the President are available at: https://www.presidencia.pt/atualidade/toda-a-atualidade/2021/03/presidente-da-republica-veta-eutanasia-por-inconstitucionalidade/

16. All details regarding this amending bill are available at: https://www.parlamento.pt/ActividadeParlamentar/Paginas/DetalheDiplomaAprovado.aspx?BID=23303

17. All details regarding the presidential veto are available at: https://app.parlamento.pt/webutils/docs/doc.pdf?path=6148523063484d36 4c793968636d356c6443397a6158526c6379395953565a4d5 a5763765247396a6457316c626e6527663306c7561574e705958527064 6d45764d444131596a6b334d4467744d7a5a694e5330304f5455324 c57466a4d324d745a6a426842e6a67325a54517a4e6a5a526d4c6e426b 5a673d3d&fich=005b9708-36b5-4956-ac3c-f0a686e4364f.pdf&Inline=true

18. All details regarding this bill are available at: https://www.parlamento.pt/ActividadeParlamentar/Paginas/DetalheIniciativa.aspx?BID=121331

19. All details regarding this bill are available at: https://www.parlamento.pt/ActividadeParlamentar/Paginas/DetalheIniciativa.aspx?BID=121467

20. All details regarding this bill are available at: https://www.parlamento.pt/ActividadeParlamentar/Paginas/DetalheIniciativa.aspx?BID=121482

21. All details regarding this bill are available at: https://www.parlamento.pt/ActividadeParlamentar/Paginas/DetalheIniciativa.aspx?BID=121545

22. All details regarding this resolution project are available at: https://www.parlamento.pt/ActividadeParlamentar/Paginas/DetalheIniciativa.aspx?BID=121485

23. All details regarding this resolution project are available at: https://www.parlamento.pt/ActividadeParlamentar/Paginas/DetalheIniciativa.aspx?BID=152124

24. This decree may be consulted here: https://www.parlamento.pt/ActividadeParlamentar/Paginas/DetalheDiplomaAprovado.aspx?BID=33774

25. All details regarding this statement from the President are available at: https://www.presidencia.pt/atualidade/toda-a-atualidade/2023/01/presidente-da-republica-submete-eutanasia-ao-tribunal-constitucional/

26. All details regarding this decision from the Constitutional Court are available at: https://www.tribunalconstitucional.pt/tc/acordaos/20230005. htmll and https://www.tribunalconstitucional.pt/tc/acordaos/20230005.htmll

27. All details regarding this Presidential veto are available at: https://www.presidencia.pt/atualidade/toda-a-atualidade/2023/01/eutanasia-presidente-da-republica-devolve-diploma-ao-parlamento/

28. All details regarding this amending bill are available at: https://www.parlamento.pt/ActividadeParlamentar/Paginas/DetalheDiplomaAprovado.aspx?BID=33892

29. The full text of the Presidential veto may be consulted at: https://app.parlamento.pt/webutils/docs/doc.pdf?path=6148523063484d364c793968636d356c6443397a6158526c63793959566b786c5a7939456 2324e31625756756447397a5357357059326c6864476c32 595339684d574d7a4d5749784e6930304e7a4a6c4c5451775a446b744 f5449305a69316b4d5467324d32526c5a6a6b344d5749756347526d&fich=a1c31b16-472e-40d9-924f-d1863def981b.pdf&Inline=true

Ayes Or Noes: Positions and Arguments used by Political Parties in the Parliamentary Debates

Abstract In this chapter, we analysed the positions and arguments used by political parties during the parliamentary debates through the six rounds. We concluded that six different political parties were in favour of legalizing active euthanasia, five of them submitting bills—Left Bloc; Green Party; Socialist Party; People-Animals-Nature Party; and the Liberal Initiative—and another party just expressing its support—the Free Party. The most used arguments were that euthanasia is an act of individual autonomy and mercy, while also mentioning that the proposed precise criteria for euthanasia acts would prevent the 'slippery slope'. They also mentioned that the discussions about palliative care and euthanasia should be separated and that no referendum should ever take place regarding human rights.

On the other hand, four political parties voted against the legalization of euthanasia—the Communist Party, the Social Democrat Party, the Popular Party and the Enough! Party. The most common arguments were the need to reinforce the supply of palliative care before any discussion on euthanasia, the respect for the right to life, or the danger of the 'slippery slope'. Moreover, they stressed the opposition from most of the consulted entities, the Constitutional Court and the President.

Keywords Euthanasia; Political parties; Parliamentary debates; Positioning; Arguments

© The Author(s), under exclusive license to Springer Nature 67
Switzerland AG 2023
I. S. Almeida, L. F. Mota, *Politics and Policies in the Debate on Euthanasia*, https://doi.org/10.1007/978-3-031-44588-0_4

As expected, political parties were deeply involved in the debate about euthanasia. As mentioned earlier, this involvement took place, first of all, through the submission of several bills by five different political parties—the Left Bloc (BE); the Green Party (PEV); the Socialist Party (PS); People-Animals-Nature (PAN) and the Liberal Initiative (IL). As well as the bills, political parties were also involved in the discussion of this topic in parliamentary debates, namely in the four plenary meetings when the mentioned bills were officially presented and voted: in the first round, on May 29, 2018[1]; in the second round, on February 20, 2020[2]; in the third round, on November 4, 2021[3]; in the fourth round, on June 9, 2022[4] and, in the fifth round, on March 31, 2023.[5]

In this chapter, we highlight the main positions and arguments in favour of and against the legalization of active euthanasia demonstrated by political parties with seats in Parliament during the mentioned six parliamentary debates. After subsections dedicated to each of the ten political parties, in which we highlight their main arguments by underlining them (Sects. 4.1 and 4.2), we summarize and compare these arguments in Sect. 4.3. To understand the depth and consequent length of each analysis, it is important to understand that the time each political party must intervene in debates depends on the number of MPs they have in Parliament.

Before moving on to the analysis of the arguments, it is important to highlight how each political party voted in the mentioned four moments. As can be seen in Fig. 4.1, where political parties are displayed from far-left (PCP) to far-right (CH), there is a wide consistency throughout the four stages of voting. On the one hand, voting in favour of legalizing euthanasia were all the MPs from the libertarian left-wing Left Bloc (BE), the environmentalist left-wing Green Party (PEV), the libertarian left-wing Free party (L), the centrist environmentalist People-Animals-Nature party (PAN) and the libertarian right-wing Liberal Initiative (IL), most of the MPs from the centre-left Socialist Party (PS) and a few MPs from the centre-right Social Democrat Party (PSD). On the other hand, voting against the decriminalization of euthanasia were all the MPs from the conservative left-wing Communist Party (PCP), the Christian Democrat and liberal Popular Party (CDS-PP) and the extreme-right populist Enough! party (CH), and most of the MPs from the centre-right Social Democrat Party (PSD).

In the following sections, we will analyse the interventions in those four debates by representatives of each of the ten parties, namely the main arguments they used against or in favour of legalizing euthanasia.

	1st round	2nd round	3rd round	4th round	5th round	6th round
Communist Party (PCP)	Against	Against	Against	Against	Against	Against
Green Party (PEV)	In favour *	In favour *	In favour *	N/A	N/A	N/A
Left Bloc (BE)	In favour *	In favour *	In favour *	In favour *	In favour *	In favour *
Free Party (L)	N/A	N/A	N/A	In favour	In favour	In favour
Socialist Party (PS)	Mostly in favour *	Mostly in favour *	Mostly in favour *	Mostly in favour *	Mostly in favour *	Mostly in favour *
Two unregistered MPs **	N/A	In favour	N/A	N/A	N/A	N/A
People-Animals-Nature party (PAN)	In favour *	In favour *	In favour *	In favour *	In favour *	In favour *
Social Democrat Party (PSD)	Mostly against	Mostly against	Mostly against	Mostly against	Mostly against	Mostly against
Liberal Initiative (IL)	N/A	In favour *	In favour *	In favour *	In favour *	In favour *
Popular Party (CDS-PP)	Against	Against	Against	N/A	N/A	N/A
Enough! (CH)	N/A	Against	Against	Against	Against	Against

Notes: * means the party was submitting a bill; ** These two unregistered MPs were former MPs from PAN and the Free Party. N/A means the political party was not represented in the Parliament at that time.

Fig. 4.1   Positions of the political parties expressed through votes throughout the six rounds. (Notes: * means the party was submitting a bill; ** These two unregistered MPs were former MPs from PAN and the Free Party. N/A means the political party was not represented in the Parliament at that time. Source: own production)

4.1   THE AYES: PARTIES IN FAVOUR OF THE DECRIMINALIZATION OF EUTHANASIA

4.1.1   Green Party

As mentioned before, the Green Party (PEV) was one of the political parties which submitted bills in rounds 1, 2 and 3, but did not submit in the following rounds as they could not elect MPs in the legislative elections 2022.

When presenting the bill from the PEV in the **first round**, MP **Heloísa Apolónia** began her speech by saying that this is a debate that deals with values which are deeply rooted in society and that, therefore, it is a delicate and sensitive issue. She clarified that no patient should be incited or advised to practise euthanasia, so as to guarantee there is no influence or pressure. She thus considered legitimate both the will of people who do not want to hasten their death and that of those who, in the same critical situation, decide to end their life. For the PEV, the State should not prohibit this possibility and must ensure respect for the principle of individual dignity, autonomy and self-determination, as people are capable and able to determine and choose what they want or what they do not want from their lives.

The PEV also considered that there should be a clear separation between the practices of euthanasia and palliative care, stating that there is a Basic Law for Palliative Care, and that *"if there is no greater investment in such care, as well as in continued care, it is because there have been, over time, those who exchange this investment for the figures of State deficit"*. Therefore, she clarified that this bill does nothing to reduce the State's responsibility to ensure patients' access to palliative care and a good supply of such care.

Heloísa Apolónia also recognized that health professionals must have their right to conscientious objection ensured. However, she considered that patients should be guided by someone who is not a conscientious objector and that euthanasia should be performed only in public hospitals to avoid private hospitals profiting from this procedure.

Finally, she confessed that she considers death to be something shocking and frightening, although these same words can also be used to define deep pain and suffering, resulting from incurable, fatal and terminal diseases, and that *"the answer will depend on each person in particular"*.

The PEV **second-round** bill was presented by MP **José Luís Ferreira**, who began his speech by saying that a sensitive subject was under discussion and that it involved fundamental values which are rooted in our society. He considered that in specific cases—in which the patient suffers from an uncured, irreversible, and fatal disease which causes intolerable suffering—euthanasia should be permitted as an act of compassion, as it allows patients not to live in such conditions and to guarantee their dignity.

The PEV MP stated the position that his party agrees with medically assisted death in extreme and well-defined situations and following an express request from the patient, making it clear that the bill should not reduce the State's duty to ensure patients' access to palliative and continued care. He added that no one will be obliged to choose to hasten their death, due to influence or pressure, nor will health professionals be obliged to monitor and assist the process if it is against their principles and convictions.

Finally, after summarizing his party's bill, the MP considered that the Parliament was able to respect the law in a thoughtful and reflected way, and, with the involvement of all members, was capable of producing a tolerant, rigorous and balanced legislation.

The conciliatory bill from the **third round** was also presented by MP **José Luís Ferreira**, who guaranteed that the conditions for completing the current legislative process are met and that the Constitutional Court did not present any reservations about the concept itself, ensuring that medically assisted death is not unconstitutional. At the end of his speech, he regretted that his party's proposal for ensuring medically <u>assisted death would be available only in public hospitals</u> did not deserve the proper attention and that *"this law will not help to reduce, lighten or excuse the State from its duty to ensure patients access to <u>palliative care and to ensure a good network of continued care</u>"*.

4.1.2   The Left Bloc

The Left Bloc (BE) was also one of the political parties that submitted bills about euthanasia in the six rounds of discussion.

When presenting the BE's bill in the **first round**, MP **José Manuel Pureza** began his speech by praising the struggle of Laura Ferreira dos Santos and João Ribeiro Santos, promoters of the 'Civic Movement for the Decriminalisation of Assisted Death' and of the petition 'Right to Die with Dignity', which set the topic of euthanasia on the political agenda. The MP from the BE thus considered that it was time to legislate the practice of euthanasia and clarified and summarized the BE's bill (see Chap. 3). Later, he clarified that the bills under discussion <u>aimed to prevent any possibility of legitimizing the early death for situations</u> other than what the bills propose.

In the **second** round, it was again MP **José Manuel Pureza** who presented the BE's bill. In his speech, he stated that the big issue was the adoption of a balanced, prudent, and rigorous law, which respected the <u>decision of each person</u> at the end of their life and provides *"(...) rich and poor the same right to have the end of life that best respects <u>their sense of dignity</u>"*, and not only to rich people who go to a foreign country to achieve it. Quoting João Semedo—a deceased former member of the BE and a strong advocate of euthanasia—he challenged the remaining MPs by saying:

> The decriminalization of assisted death is the <u>most humanitarian and democratic option</u> we can approve for the end of life: no one is obliged to do it, and no one is prevented from doing it, the only criterion being each one's choice. After all, isn't this democracy?

He also considered that the decision of all MPs should not follow *"opportunistic political manoeuvres, nor emotional blackmail"*, and that an eventual referendum would be a *"cynical use of the democratic instrument"*.

At the end of his speech, after summarizing his party's project, the MP reported cases of people with whom he had spoken about rights and dignity at the end of life. Finally, he asked the following question: *"Have we chosen the arrogance to impose on everyone an end-of-life model that means unbearable violence for many or, refusing any imposition, have we decided to respect the choice of each one over the end of his life?"*.

During the debate, MP **Pedro Filipe Soares** also intervened, mentioning that the debate revealed two ideas: The first is the awareness that all citizens now have about euthanasia; and the second is the absence of social alarm after such a wide discussion of the topic. During his speech, he shared several examples of personalities, such as José Saramago—a famous writer and historic member of the Communist Party—, who was in favour of euthanasia, or Ramón Sampedro, who in 1968 became a quadriplegic and fought more than three decades for the right to end his life.

He also stated that two lessons can be learned: The recognition that our dignity is defined by us and that we must have a law that gives us freedom of choice. In this sense, he stated that the distinction between palliative care and assisted death diverted the debate from the essential since palliative care does not solve all problems and science itself has limits. On the other hand, regarding the argument of the 'slippery slope', he mentioned that opponents said the same when the laws on abortion and drug use were changed. At the end of his speech, he mentioned that all the conditions for a valid request *"(...) are attested by medical professionals and, not less important, the decision to take the procedure of anticipating death to the end is always exclusively and uniquely from the patient and revocable at any moment"*.

The conciliatory bill from the **third round** was also presented by MP **José Manuel Pureza**, who stated that *"the Constitutional Court declared, unequivocally, that there is no unconstitutionality of principle"*. He added that the court stated that the right to life should not impose a duty to live and it was, therefore, legitimate for Parliament to decriminalize assisted death in very limited circumstances.

The bill of the BE in the **fourth round** was presented by MP **Catarina Martins**, who recalled the words of João Semedo: *"The decriminalization of assisted death is the most humanitarian and democratic option that we can approve for the end of life: no one is forced, and no one is stopped from* *doing so, the only criterion is the choice of each one"*. She added that it was the third time the issue was under discussion in Parliament and that the President has used his veto power twice, stating that his second decision was a personal one.

She added that euthanasia is <u>the most debated topic in 48 years of democracy</u> in a discussion that has been extended to the whole of society. Finally, she mentioned that the argument of the slippery slope was also discussed regarding abortion and the reality demonstrated that the number of abortions has been decreasing.

During the **fifth round**, regarding the reconsideration of the decree, MP **Catarina Martins** claimed that the changes introduced in that amending bill were <u>in line with the requests/suggestions made by the Constitutional Court</u> and that there is a <u>huge consensus about euthanasia</u> in Portuguese society.

Finally, during the **sixth round**, regarding the reconsideration of the decree, MP **Catarina Martins** stated that the doubts raised by the President of the Republic were only about regulatory details, which meant that the bill should be approved. She recalled that this approval would be a decisive step in society becoming more respectful of everyone's will. At the end of her speech, she accused the parties against medically assisted death of opportunism for wanting a new review from the Constitutional Court.

4.1.3   The Socialist Party

Like the BE, the Socialist Party was also one of the political parties that submitted bills about euthanasia throughout the six rounds of discussion.

The bill of the PS in the **first round** was presented by MP **Maria Antónia Almeida Santos**, who began by saying that although *"living is a right and protecting life is a duty of the State (...) we <u>cannot ignore the people for whom the diagnosis is irreversible and who are in extreme suffering</u> (...)"*. In this regard, she stated that the Constitution defines human life as inviolable, *"but <u>not as an indispensable duty</u>"* and that legislating on the conditions under which euthanasia is not punishable meets the request of sick people themselves.

For this reason, the MP stated that euthanasia will not be practised if patients are not conscious and duly informed, there are doubts about patients' judgement, the opinion of the guiding physician and the specialist physician is unfavourable, or patients do not reiterate their will.

In the **second round**, the PS bill was presented by MP **Isabel Moreira**, who began her intervention by saying that no life is more dignified than another and that the PS is honoured in its tradition of making criminal policy decisions based on individual autonomy, without violation of the Constitution, so that there is room for individual intimate choices because it concerns only people themselves. She added that the State cannot impose a single path of individual choices nor a single conception of life and would be a totalitarian State if it did so.

The MP also said that it was revolting when she heard that the proposed bill was not necessary and that people could always commit suicide, judging this statement to be cruel, and also giving the example of patients who ask for help to die in the toilet or who go abroad for that purpose. She also quoted João Semedo when he said *"helping to die peacefully, ending useless suffering, is an attitude of high moral value and great humanism"*. She ended her speech by responding to those who named the socialists radicals, by saying that radical was *"(...) to say suffer, suffer until your heart stops beating, about while you are a body, because for us this is how people should die"*.

Still representing the PS in this debate, MP **Alexandre Quintanilha** began his speech by mentioning the current Living Will, which allows patients to request the possibility of avoiding the artificial extension of their life but does not help them to finish their lives. The bill was therefore oriented towards the values of courage and humanity, respect and dignity. At the end of his speech, while responding to the arguments of the 'slippery slope', he mentioned the successes in the issues of abortion and drug use.

The conciliatory text of the **third round** was also presented by MP **Isabel Moreira**. She considered that MPs were in a moment of reappreciation of the bill after analysis of the President and the Constitutional Court, in which the latter went beyond the received request and made sure to specify that euthanasia was not unconstitutional for violation of the principle of human life (see Chap. 3, Sect. 3.2.8). In this regard, the MP mentioned that the court was only unsure about the concept of "definitive lesion of extreme gravity", since, unlike the President, it considered the concept of "intolerable suffering" to be determinable. Finally, in response to criticisms regarding the rush in approving this bill and that no entities had been involved in hearings, she mentioned that there was no need to do so, considering this bill emerged as a response to the decision of the Constitutional Court.

The **fourth-round bill** was also presented by MP **Isabel Moreira** who started her speech by recalling that this <u>debate had been brought on for more than 7 years</u> by several prominent people, such as Laura Ferreira dos Santos and João Semedo. She also raised some doubts about the second veto from the President, which was political in nature and related to the objection on the final draft of the bill performed by the CDS-PP. After these initial considerations, the MP continued by saying that the bill was in line with what had been previously proposed, which was based on a thorough process of comparative analysis of the law. She finished her speech by mentioning examples of people who are waiting for euthanasia to be decriminalized to make their requests.

During the debate, MP **Alexandre Quintanilha** said that it was time to complete this extensive process and made the following remarks:

Many realize that the additional years of life that medicine provides are not always accompanied by the promised and desired quality and are clearly aware that the loss of autonomy, self-esteem and dignity, as well as the physical and psychological suffering they feel, will increase in the time of life they still have.

He then reflected on several criticisms raised against the decriminalization of euthanasia, stating, for instance, that *"the additional argument that assisted death saves money for the State is as unfair as that of palliative care being a huge source of revenue for professionals and health institutions"*. At the end of his speech, he then questioned whether it would not be more honest and coherent to end the criminal penalty of assisting others in dying, and that democracy gives us the right to choose how we want to live and the duty to respect the right of others.

In another perspective from the PS, MP **Alexandra Leitão** said that this is an issue at the heart of the powers and legitimacy of a democratic and representative State and criticized the idea of organizing a referendum.

To finish the debate, MP **Eurico Dias** said that the legislative initiative of his party derived from a <u>long work with the participation of other political parties and civil society</u>. He then added that the bill was *" (…) part of the <u>framework of human rights, individual freedom and the right to a dignified life,</u> which should not be subject of a referendum"*. He finally stated that if we have the right to life, it is no less true that one should defend <u>life with dignity</u> and that it should not be confused with a duty to live.

During the **fifth round**, MP **Isabel Moreira** mentioned that the proposed bill has been one of the most <u>debated and scrutinized</u> in the political history of Portugal, which makes it a quite balanced document. She also claimed that the changes introduced in the amending bill were a result of requests/impositions from the Constitutional Court. Finally, she referred to several cases of citizens that have been asking for euthanasia and several activists who advocated for this law to be approved.

During the **sixth round**, regarding the reconsideration of the decree, MP Isabel Moreira reaffirmed that this was the moment to respect those who longed for this day, in which medically assisted death is consecrated after such a long time. She also recalled that this was not a matter of conscience, but of criminal policy and <u>respect for human rights</u>, giving examples of successive international courts when they were called upon to verify the constitutionality of this issue.

4.1.4   The People-Animals-Nature Party

Like the BE and the PS, PAN also submitted bills during the six rounds. This party was even the first political party to do so in the first round, almost one year before the following three political parties.

The bill from PAN in the **first round** was presented by MP **André Silva**, who began his speech by congratulating the 'Civic Movement for the Decriminalisation of Assisted Death' for the proposed petition and the quality, seriousness and elevation of the discussion held in the committee. He then mentioned the main goal of the bill (see Chap. 3) and that "*(…) to prevent voluntary anticipation of death on request in very special contexts is to prevent an <u>act of altruism</u>*". Later, André Silva added that <u>not all suffering is treatable</u> and that palliative care is not the only solution for all cases. The PAN MP also considered that, currently, patients are prevented from deciding about their own life and that their <u>autonomy was conditioned</u> by legal impositions, stating that a failure in autonomy at the end of life is like failing a whole life.

At the end of his speech, he stated that the Constitution determines human life as inviolable but also mentions as a fundamental right "*(…) the free development of personality as the right to make a plan of life, as well as the freedoms of conscience and thought and dignity of the human person*". In this sense, the irreversible decision that assisted death entails was intertwined with <u>guarantees of formal and secure processes</u>, which are scrutinized and regulated.

In the **second round**, the bill from PAN was also presented by MP **André Silva**, who started his speech by mentioning that the debate was about empathy, solidarity and the ability to put ourselves in the place of others and about the courage and responsibility to build a fair law that can respect the will and decision of each person. He then considered that the State was not enabling freedom of choice. In this context, he asked MPs the following question:

> Is it acceptable that a rich Portuguese person can guarantee his/her will to anticipate death by leaving the country, and that this same country does not give the universal response that is imposed in a democratic and plural rule of law, that is, the conditions for this will to be respected here, in the company of family and friends?

He also considered that those who argued that the discussion remained undone and that the decision was being made 'in a hurry' were hiding the real goal of their statements, which was to prevent the decriminalization of euthanasia. In this regard, and concerning the possibility of a referendum, he stated that this proposal had a political purpose and that the referendum could not be trivialized as *"the last stronghold of those who want to stop the extension of people's autonomy and self-determination"*.

At the end of his speech, after summarizing the party's bill, he stated the position that the decision to decriminalize medically assisted death is final, with risks and irreversible effects, but argued that the bills ensure the formality and safety of cases through enormous scrutiny and regulation. In this regard, he mentioned the opinion about their first-round bill expressed by the unsuspicious National Council of Ethics for Life Sciences, which considered that the *"procedure provided by the bill from the PAN has the merit of being able to reduce the risk of requests that do not correspond to a genuine and consistent use of the will itself, and also to limit bad judgments on the conditions assumed by law enforcement"*.

The perspective from PAN was also defended by MP **Bebiana Cunha**, who declared that her party was on the side of all patients, family members and professionals who hoped that medically assisted death could be possible. She also mentioned that several constitutionalists consider that legislation decriminalizing medically assisted death is constitutional. At the end of her speech, she stressed that her party's bill was not against the professional practice of those whose mission is to save lives, but it wanted all possibilities to be taken into consideration.

PAN was represented in the presentation of the conciliatory text of the **third round** by MP **Bebiana Cunha**, who said that we were facing a path of human dignity, with a debate extended to the entire civil society and that justice serves and defends those who suffer, those who have no cure and that they have the right to make decisions about their own life. She also stated that the Constitutional Court considered that *"(...) the inviolability of human life enshrined in the Constitution does not constitute an insurmountable obstacle to decriminalizing the anticipation of medically assisted death under certain conditions"*. At the end of her speech, she stated that the decision could not continue to be postponed and that there was no possible justification for the injustice that these successive postponements bring to human dignity and the suffering of people.

In the **fourth round**, the bill from PAN was presented by **Inês Sousa Real**, who began her speech by saying that *"the issues of life and death are complex, even more so when it comes to someone who has the misfortune of finding themselves in a situation of serious and incurable illness, which causes them irreversible and intolerable suffering"*. She then stated that this was one of the most discussed and reasoned debates, in which civil society associations and entities were heard, and constitutional certainty was ensured by the judgment of the Constitutional Court.

During the **fifth round**, MP **Inês Sousa Real** stressed the existence of a great debate about euthanasia and the arguments that its decriminalization would be a way to respect citizens' right to choose and human dignity.

During the **sixth round**, regarding the reconsideration of the decree, MP **Inês Sousa Real** said that this was a debate on the respect for human dignity and that this had been a wide-ranging debate with several parliamentary hearings, which thus meant that it was a text with all the constitutional conditions to be promulgated and that it could be properly applied in the future.

4.1.5   Liberal Initiative

The Liberal Initiative political party also submitted several bills, in all rounds except the first, as it did not have parliamentary representation during that time.

During the **second round**, the bill from the IL was presented by its only MP **João Cotrim de Figueiredo**, who began his intervention by mentioning that euthanasia was not an easy topic. In this sense, he ensured that the bill from his party respected the <u>express, free, current and enlightened will</u> of the patient.

In the **third round**, the conciliatory bill was also presented by MP **João Cotrim de Figueiredo**, who began his intervention by claiming that this moment was the end of a process that led to a debate with <u>thorough participation in Portuguese society</u>, adding that no one could deny the sensitivity with which the subject had been addressed. At the end of his speech, he then took the position that *"(...) the <u>decision of the end must be ours</u>, because life, from the beginning to the end, is what we believe and what we do, freely, informedly, and consciously"*.

In the **fourth round**, the bill from the IL was also presented by **João Cotrim de Figueredo**, who said that it was the third time the topic was under discussion, and it would be strange if new arguments or new considerations were presented. Nevertheless, he considered it was a good time to reaffirm principles, one of them being that the decriminalization of assisted death is not the same as defending its actual use.

Regarding the discussion process, he stated that the party would vote in favour of all bills, but it would vote against the proposal of a <u>referendum since it was a 'clumsy' attempt to set a precedent that fundamental rights and individual freedoms should be subjected to referendums</u> and to make us believe that direct democracy incorporated virtues unattainable by representative democracy.

During the **fifth round**, MP **Patrícia Gilvaz** argued that there had been a lot of debate about euthanasia, which led to a wide consensus on the topic. Moreover, she stressed that the right to live should not lead to an obligation to live and that <u>individual choices should prevail over collective moral</u>. She finished her statement praising the value of representative democracy, probably to answer MP André Ventura's previous claim for a referendum.

During the **sixth round**, regarding the reconsideration of the decree, MP **João Cotrim de Figueiredo** mentioned that the messages of the President and the Constitutional Court had been respected and considered over time, so he did not agree with the argument that the bill was an 'affront' to the President. The MP demonstrated, again, that his party was against the referendum on this issue and challenged the parties that mentioned the need to request a review of constitutionality to do so.

4.1.6   The Free Party

The only party which voted in favour of bills but did not submit one was the Free party—which had been able to elect one MP in 2019, who later became an unregistered MP before the discussion of the bills and elected another MP in 2022. Therefore, the party started to participate in the debate only in the **fourth round** of discussion, represented by MP **Rui Tavares**, who said that *"euthanasia, a good death, proper death, healthy death is equivalent to eudemonia, to the proper, good and healthy life"*. He also mentioned that the same philosophy which invented the words 'politics' or 'democracy' gave us the responsibility of ensuring the principles and conditions of <u>autonomy to make conscious, dignified and autonomous decisions</u>. At the end of his intervention, he said that his party had decided not to present a bill because the remaining bills were already very solid and similar but would participate in the discussion to be held in the specialized committee.

During the **fifth round**, MP **Rui Tavares** stressed that decriminalizing euthanasia would be a way to respect <u>freedom of choice</u> and the rule of law.

Finally, during the **sixth** round, regarding the reconsideration of the decree, MP **Rui Tavares** mentioned that the presidential veto had always divided Parliament and that it was only the result of the normal functioning of institutions. On medically assisted death, the MP said that it was about <u>respecting what each one decided</u> about their death and that the referendum was not the place for majorities to decide on the individual rights of potential minorities, hoping that people asking for euthanasia due to their suffering would always be a minority.

4.2   THE NOES: PARTIES AGAINST THE DECRIMINALIZATION OF EUTHANASIA

4.2.1   The Communist Party

As stated earlier, the Communist Party (PCP) decided to vote against all bills throughout the six rounds of discussion.

In the **first round**, the PCP was represented by MP **António Filipe** who considered the debate about euthanasia should not be summarized as being about individual conscience, but a rather complex State decision with <u>profound social consequences</u>. He then asserted that the <u>Constitution gave the State the obligation to defend a dignified life</u> for its citizens and the situations which lead people to submit a request for assisted death should be considered, by saying the following words:

> (...) people who wish to anticipate the end of their life because the necessary care for an end of life without suffering is not guaranteed, or because they are refused the material means to have an end of life in conditions of dignity, or because the necessary support in the lack of family support is not guaranteed only deserves understanding, solidarity, and support so that they have a real alternative.

In this sense, he said that the PCP would continue to strive for the implementation of a strengthened National Health Service that guarantees all the needs of its users, giving the example that science had resources that allow physical and psychological suffering to be reduced or eliminated, and also mentioned the existence of legal documents to ensure the individual will, such as the Living Will. He also expressed his concern about the 'slippery slope', considering that, once euthanasia was approved, it would be used in cases that were not foreseen by the bills, mentioning Dutch statistics as an example. In addition, he said that in countries where the practice was legal, euthanasia had become an "(...) international business (...) [and that] the nature of capitalism takes care of making it much easier".

Finally, he mentioned that none of the 2015 manifestos for legislative elections included anticipated death and reported that the PCP had decided to vote against all the bills under discussion and that the party's position was not based on moral or religious considerations, but rather on the idea that the intrinsic value of life should prevail, beyond its usefulness, or debatable economic and social standards.

In the **second round**, the PCP was once again represented by MP **António Filipe,** who reiterated the idea that the debate about euthanasia should not be based on individual consciences, but on the reflection about the attitude which the State should take towards the terminal phase of the lives of its citizens. In this sense, he stated the following:

> Individual autonomy is something that must be respected, but an organized society is not a mere sum of individual autonomies. The legislature cannot assume a legislative choice about people's lives and deaths without taking into account the circumstances and social consequences of that option.

He then stated that individual dignity was not under discussion, unlike what happens with the State, which denied many citizens the means to live with dignity. In this regard, he mentioned that <u>palliative care in Portugal was only accessible to 25% of the population</u>, which meant the State should join forces to create a network of palliative care and health care with a universal nature.

At the end of his speech, he demonstrated the contradiction of bills, by saying that if death were a right, it would not be lawful to defend the anticipation of death to be dependent on the decision of third parties. Finally, he mentioned that the issue of euthanasia had been discussed around the world for many years and that it was no accident that most countries had rejected it, and that he regarded the <u>concern about the situation of the few countries where euthanasia was legal</u> as something that could not be ignored.

The PCP was represented in the discussion of the **third round** once again by MP **António Filipe**, who mentioned the bill which his party voted against had been declared unconstitutional and that it would not be possible to know whether the new terms and concepts would be sufficient to resolve the issues that had been raised by the court. At the end of his speech, the MP stated that the State could not deny the provision of health care to its citizens and that the <u>creation of a palliative care network should be the priority</u>, adding that the evolution of science and technology had allowed advances in medicine.

In the **fourth round** of discussion, the PCP was represented by MP **Alma Rivera**, who began her speech by saying that it was the third consecutive legislature in which legislative initiatives on assisted dying were discussed and that bills had been vetoed twice. She then stated that her party refused to view this debate as a 'trench war' between religions and atheisms and that it was about a <u>State option</u> rather than a judgement on individual choices.

She therefore declared that the State must strive to mobilize its resources and scientific knowledge, rather than to create conditions to anticipate death and that no one understands euthanasia as a substitute for palliative care. On the topic of the referendum, she said that the PCP considered that legislative options on fundamental rights should not be subject to contingencies, Manichaeism and the simplification of popular consultation, adding that Parliament had the necessary legitimacy to decide.

During the **fifth round**, MP **Alma Rivera** claimed that society should not be a sum of individual liberties and that one should take into consideration the context where a bill is being discussed, since euthanasia may be dangerous given the lack of conditions and State support regarding living conditions, in general, and healthcare provision, in particular. She also rejected the idea that the value of human life should be overpowered by liberal considerations about the ability to work or the costs of an ill person.

During the **sixth round**, regarding the reconsideration of the decree, MP **Alma Rivera** reaffirmed that her party's vote considered the values in question and that it was not a war of religions, and recalled, therefore, that it was an option of the State, which can no longer deny the majority of citizens the health care they need in times of greatest suffering. In the end, she stated that the dignity of the human person was not under discussion, but whether a State that denies health care should create mechanisms for death.

4.2.2   The Social Democrat Party

As explained earlier, the centre-right Social Democrat Party gave a 'free vote' to its MPs during the six rounds of discussion, which led to most MPs voting against the proposed bills, although a small group of MPs had voted in favour of them, particularly the ones presented by the Socialist Party. In some rounds of discussion, this party was therefore represented by MPs who voted differently.

Regarding the **first round** of discussions, the PSD was firstly represented by MP **Fernando Negrão**, who voted against the bills and believed that the opposition to the decriminalization of euthanasia in the circumstances described was the majority feeling of his political party, considering that the discussion of the topic took place *"(…) without sufficient debate and consideration, without demanding rigorous, exhaustive and cautious reflection"*. During his speech, the MP implied that political parties lacked some legitimacy to submit bills considering they had not mentioned the issue of euthanasia in their manifestos for the 2015 legislative elections.

In addition, he argued that 80% of Portuguese patients die without access to palliative care, that is, without any aid in pain relief in the terminal phase of their life. In this regard, he then asked whether progress was based on the *"vanguardism of being the 4th country in Europe to approve the decriminalization of euthanasia (…) or is it rather led by the fact that we are at the tail end of Europe concerning palliative care?"*.

Also participating in this first round of discussion was MP **Margarida Mano**, who also voted against the bills and began her intervention by stating that, since we live in a society that imprints social changes and involves ethical dilemmas and the defence and guarantee of individual freedoms, the individual good and the common good will always be two inseparable variables of the political equation of legislators.

Another argument raised by this MP was the fact that the <u>CNECV issued an opinion</u> on the PAN project, <u>in which it expressed concerns about the conditions of inequality</u> that would be created concerning the health-care due to citizens. Moreover, she pointed out that the <u>Order of Physicians was against all bills</u>. In political terms, she questioned if the moment for this discussion was opportune, as <u>no party had expressed its positions on euthanasia in their manifestos</u>.

In the **second round** of discussion, the PSD was represented by MP **António Ventura**, who began his intervention by praising his party for giving them a 'free vote'. Having said that, he asserted the following:

> *The experience of other countries and several studies have told us that often when someone asks for death, it can result from an <u>occasional awareness</u>, likely to be overcome, or from <u>depressive states</u>, which can be treated. In other words, it is never certain that one reliably respects the will of the person who asks for euthanasia.*

In this regard, he mentioned that in Portugal, where there is still a lot of work to be done in offering health care and palliative care—only about 25% of the population has access to it—and where there is a lack of human and financial resources in the National Health Service, the solution should not be the provision of euthanasia. Therefore, he thought that once approved, euthanasia "would gain wings", moving rapidly and at any time from cases of terminal illness to chronic disease, disability, to children, in a process named the 'slippery slope', giving the examples of the Netherlands and Belgium. He ended his intervention by recalling that the bills received negative opinions from the Order of Physicians, the Order of Nurses and the National Council of Ethics for Life Sciences.

Also representing the PSD in this round, MP **André Coelho Lima**, who did not vote against the bills, began his intervention by pointing out the existence of different perspectives, which should be debated with moderation and without radicalism.

After mentioning two examples of physicians who are in favour of legalizing euthanasia—João Ribeiro Santos, former director of Hospital Curry Cabral, and his father—he assumed the central question as being one of allowing the holders of life by right to decide on their death. At the end of his speech, he claimed to be against a State that can decide with and under what conditions someone should die is also to be against a State that decides to keep us alive even against our will. He concluded by saying the following sentence:

> *I am in favour of life, I am in favour of dignity, I am in favour of self-determination, I am in favour of love for one's neighbour, I am in favour of a State that does not want to moralize society, I am in favour of the principle of non-interference in the lives of each one of us. I am all of this because I believe that every human being—and their life—should be respected by everyone to the point that we, the others, do not want to have the pretension of deciding on it.*

Later, MP **Sofia Matos**, who also voted in favour of the bills, began her speech by praising the existence of different opinions in Parliament and in her party. She also mentioned that the regulation of euthanasia could not mean less State investment in the palliative care network, which she considered insufficient and a priority. She then concluded by stating she would vote in favour, thus *"giving these citizens the possibility of choosing to live or die, according to the criteria of dignity that each of them has built throughout their lives"*.

In another speech, MP **Cláudia Bento** stated that this is an issue that needs reflection since it violates one of the pillars of fundamental rights: the right to life. She then added that the legal system Portuguese already provided for several instruments related to individual freedom, such as the Living Will, also saying that there will never be a guarantee that the request for euthanasia is truly free and unequivocal since it results from a depressive state or a call for attention.

Regarding the argument of liberation from suffering, she replied that the solution should be the elimination of therapeutic obstinacy and measures to control pain and combat symptoms. In this regard, she stated that:

> *(…) it is not acceptable to place euthanasia as a therapeutic option when the National Health Service is so fragile. Moreover: euthanasia is not a medical act; it is, indeed, a violation of medical ethics. With the legalization of euthanasia, consequences have already been seen in the countries that have legalized it, the first being the destruction of the doctor-patient relationship.*

At the end of her speech, she presented the example of countries where euthanasia is legal, such as the Netherlands, and questioned what could be considered a "definitive injury", taking as an example being deaf or having an amputation. She, therefore, considered that Parliament should avoid legislating in 'haste'.

In the **third round** of discussion, PSD was represented by MP **Adão Silva**, who had a neutral vote and declared that he would not be swayed by the normal controversy of whether the judgement of the Constitutional Court is unconstitutional or not. Nevertheless, he considered that legislating on life-and-death issues is the supreme challenge of any legislator, which means that the necessary conditions to continue the debate were not present, given the eminent dissolution of Parliament.

The PSD was represented in the **fourth round** of discussion first by MP **Paulo Oliveira**, who started his intervention by questioning whether assisted death was the only solution. He also mentioned the <u>negative opinions from the Order of Physicians and the Order of Nurses</u> to support his vote against the bills. He finished his intervention saying he would be in favour of organizing a referendum.

This round of discussions also had the participation of MP **André Coelho Lima**, who voted in favour of the bills and started his intervention by thanking his party for giving a free vote. He then justified his opinion with the following sentence: *"I am not in favour of euthanasia, but I am not against euthanasia either, just because I understand that <u>it does not concern me what everyone chooses to do with their own life</u>. This is the structural assumption of my thinking".*

Later, MP **Cláudia Bento**, who voted against the bills, said that the previous legislature was marked by two failed attempts to approve the decriminalization of assisted death, mentioning the decision from the Constitutional Court or the <u>negative opinions from the CNECV and the professional Orders</u>. She then added that, since they were in a new parliamentary term, the new process should be started without haste. In the end, she stated that <u>Portugal is far from guaranteeing universal access to palliative care</u>, and this should be the urgency, adding her concerns about the <u>generalization of cases in Belgium and the Netherlands</u>.

Finally, MP **Sofia Matos (PSD)**, who voted in favour of the bills, stated that all the bills *"have a very specific recipient in common, <u>offer security and provide for the specific circumstances</u>* in which medically assisted and *non-punishable death can be used"*. She also said that her vote is based on the idea of giving citizens the possibility of choosing to live or die according to their criteria of dignity and that she felt unable to decide on the lives of other citizens. In this sense, she said she would never vote in favour of the referendum proposed by the Enough! party, which she considered to be insidious and would not enlighten anyone.

In the **fifth round**, MP Paula Cardoso stated that the new bill, which considered assisted suicide as a preferred method, would clash with the Penal Code, namely the article which forbids people to incite or support suicidal acts. She also claimed that the topic of euthanasia was not as consensual in Portuguese society as other MPs were claiming and that there was a need to pay attention to the context, particularly in a country that does not provide enough support to elderly citizens and healthcare provision. She also mentioned that her party was in favour of a referendum.

During the **sixth round**, regarding the reconsideration of the decree, MP **Paula Cardoso** recalled that this version had gone further than the previous ones that it was a version that changed the relationship between assisted suicide and euthanasia, and that it was no longer a choice, and the patient was forced to commit suicide. In this sense, she considered that the doubts raised by the President on the lack of definition of who would define the physical incapacity of the patient were legitimate. She then reaffirmed that the Portuguese citizens should be called upon to express their opinion in a referendum. She finally stressed that we live in a country where the elderly and fragile do not have access to physicians or continued care. At the end of her speech, she admitted that a group of PSD MPs would request a successive review of the constitutionality of the law.

4.2.3 The Popular Party

CDS-PP voted against the bills in the first three rounds, although not expressing a similar vote in the following rounds because it lost all its MPs in the legislative elections of 2022.

In the **first round**, the CDS-PP was represented by MP **Isabel Galriça Neto**, who began her intervention by stating that her party was against euthanasia, advocating the protection of life and the promotion of dignity and freedom, as well as a dignified life and a dignified death through the provision of good health-care. She added that euthanasia does not confer dignity on life or death and that *"dignity is an intrinsic value of the human being and, regardless of circumstances, there are no lives worth living and others not worth living"*. In this regard, she also said that it is absurd to speak of the so-called 'right to die' and that death is an inevitability.

During her speech, she also mentioned the issue of the 'slippery slope' by asking the following question:

> *(…) why do we ignore the reality of the few countries where euthanasia is legalized, with laws like those we debate today, such as the Netherlands, Belgium, or Canada, where people with mental illness, people in mourning, people who are not at the end of their lives, newly diagnosed people or people with disabilities are euthanized?! Why, (…) do we repeat the mistake?!.*

At the end of her speech, she said that a patient's dignity and freedom of life are ensured by widespread access to better palliative care and social support. Furthermore, she considered the debate about euthanasia to be precipitous. Finally, she clarified that the CDS-PP were involved in the discussion and approval of legislative acts such as the Law for Palliative Care, the Law of the Living Will and the legislation on the rights of sick people at the end of life and support for caregivers.

In the **second round**, the CDS-PP was represented by MP **Telmo Correia**, who began his speech by claiming that this debate was not supported by fundamental studies and fundamental elements or an assessment of the impact and consequences on the National Health Service. He accused the political parties of not including the issue of euthanasia on their manifestos for the 2019 legislative elections and assumed the position of the CDS as a humanist and personalist party and therefore assumed that no human lives are more or less dignified and that life always has the same value, from its beginning to its natural end.

Likewise, he considered that euthanasia poses a problem for health professionals, who should approach patients to help and heal them, and not to kill them since euthanasia is neither a treatment nor a medical act, and this means a breach of trust between physician and patient and in the National Health Service. He finished his speech with several questions addressed to MPs related to how to match "intolerable suffering" with an actual free and lucid choice, how the value of pain and suffering is defined, and whether the risk was being run of the 'slippery slope'.

During the **third round**, the CDS-PP was also represented by MP **Telmo Correia,** who started by saying that the MPs from his party were also voting freely, even if they were all voting against the bill. He supported this position with the following sentence:

> *Because our party has always had this position: we are against euthanasia as a matter of principle and values, we are against euthanasia because we are today, as we have always been, defenders of the supreme value of human life. From our point of view, this bill corresponds, moreover, to what could be called a death drive, and the doubts that should have been resolved in this final text were not.*

He also mentioned that Parliament was about to be dissolved and that the new text had been made in haste and secretly, without even consulting the National Council of Ethics for Life Sciences or the professional Orders. Likewise, he mentioned that the proponents of this new bill did not explain how the problems identified by the Constitutional Court and the President had been amended.

4.2.4   *The Enough! Party*

The Enough! party voted against the bills in all rounds except the first, as it only started to have parliamentary representation in 2019.

In the discussion of the **second round**, the sole MP of **CH, André Ventura,** began his speech by mentioning that the decriminalization of euthanasia was discussed exactly 100 years after the death of Jacinta Marto, seer of Fátima, and considered this a historical provocation, also mentioning that, in 1939, Adolf Hitler had decriminalized euthanasia in Germany.

He then added that it made no sense to legalize euthanasia when the country does not provide access to palliative care and Parliament refuses to criminalize families who abandon the elderly. At the end of his speech and not believing Parliament would revert its decision, he asked the President, who would have to face elections for a second term in a near time, to organize a referendum.

CH was also represented by MP **André Ventura** in the **third round**, who said it was shameful to be discussing the issue when Parliament was about to be dissolved, mentioning that the rush was a sign of the fear that a new Parliament would not approve euthanasia. At the end of his speech, he questioned the other MPs about the promised palliative care beds and the reinforcing end-of-life medical care.

In the **fourth round** of discussions, CH was also represented by MP **André Ventura**, who began his speech by stating that *"the obsession of this Parliament and the obsession of this left-wing with death is very strange"*. In this context, he said that in an extremely aged country, where 70% of Portuguese people do not have access to palliative care, the concern of left-wing parties is to give them death directly. Moreover, he mentioned that the legislative process of these new bills did not include hearings with the entities that should be heard about the topic.

He then added that this was not the way to legislate on life-and-death issues and fundamental issues, and since it is a matter of conscience that mobilizes civil society, the Portuguese population should be heard through a referendum. In this regard, he mentioned that the leader of the IL was previously in favour of holding a referendum, as were the historical members of the PSD Aníbal Cavaco Silva, Marques Mendes, Paulo Rangel and Luís Montenegro, the latter being the new leader of the PSD. He ended his speech by quoting Pope John Paul II, accusing the Parliament of finding this quote a joke, *"nothing will ever be incurable that we cannot take care of"*.

Later, MP **Filipe Melo** began his intervention by stating that the left wing of Parliament had a weird obsession with death. He also stated that their bills violated the Constitution since this document considers human dignity as inviolable and that no death penalty should exist, and also stated that he considered euthanasia to be a death penalty.

He proceeded by saying that the electoral manifesto from the PS did not cover the topic of euthanasia, which meant the bill from this party should be considered deeply illegitimate and undemocratic. He also stressed that the negative opinions from various professional Orders, such as the Order of Physicians, as well as from the civil society, the Constitutional Court and the President, had been largely ignored.

In another perspective from **CH**, **Pedro Frazão** said that *"euthanasia, which is always presented as a freedom of the people, ends up becoming a right for States to suggest death, at a time when we are not able to exercise that* *freedom or to assert our will".* He also stated it was claimed that the bills under analysis would be applied only in exceptional situations, but that they propose it to all citizens with a definitive injury of extreme severity or serious and incurable disease, which may include several diseases such as Alzheimer's or Parkinson, among others.

To finish his intervention, he pointed out that <u>what the Portuguese want is more palliative care</u>, fewer waiting lines in medical appointments and emergencies, and a family physician, while also accusing the Socialists of being afraid to listen to the Portuguese people, who want life and do not want euthanasia.

During the **fifth round**, MP **André Ventura** started his intervention by saying he did not understand why the left-wing parties had such a <u>rush and obsession</u> about legalizing euthanasia. He also stated that the introduced changes were significant, and a new bill had been created, rather than an amended one. He also claimed that no country had similar legislation and that the bill <u>does not specify who decides what procedure to be used</u>—assisted suicide or active euthanasia. He also claimed that one should take into consideration the context of a country where there is <u>not enough palliative care and where the public healthcare provision does not function</u>. He finally called on the President or the Constitutional Court to reject the bill.

During the **sixth round**, regarding the reconsideration of the decree, MP **André Ventura** said that Parliament had opened an institutional conflict with the President since the latter had raised legitimate doubts about the law, which he considered a 'legal aberration' and 'nonsense'. The MP stated that the law would never come into force, and that *"(...) to tell people that they can resort to euthanasia if they fail to commit suicide is legislative nonsense, immoral and an absolute political aberration".* He then mentioned that if euthanasia was to be legalized, a new future majority of right-wing political parties would revoke the law in the near future. At the end of his speech, the MP assured that dodging <u>the referendum was dodging democracy</u> and recalled that the voluntary interruption of pregnancy had had two referendums. He finished by asking which doctors would implement euthanasia and guaranteed that there would be none due to insecurity and that Portugal should deal with the lack of palliative care rather than with euthanasia.

4.3   A COMPARISON OF THE POSITIONS AND ARGUMENTS USED BY POLITICAL PARTIES

As can be seen throughout Sects. 4.1 and 4.2, there were two different groups of parties—those which were in favour of decriminalizing euthanasia and those against it—which advocated distinct arguments. The Fig. 4.2 below presents a summary of the positions and arguments on euthanasia from all the political parties through the six rounds.

On the one hand, the most often used arguments in favour of legalizing euthanasia were the defence of the freedom of choice and that it would give patients the possibility to avoid unbearable pain and suffering, and thus have a more dignified death. Other arguments used are the fact that the Constitutional Court declared that euthanasia, as a whole, was not against the Constitution, or that the bill was highly debated and scrutinized and was in line with the request/criticisms of the President and the Constitutional Court.

Conversely, the most often used arguments against the decriminalization of euthanasia were the need to strengthen the National Health Service and the supply of palliative care before legalizing euthanasia, or the danger of the so-called 'slippery slope', with less restrictive conditions being approved in the near future. Moreover, the objectors to euthanasia claim that there is a need to respect the right to life mentioned in the Portuguese Constitution and that several external entities expressed their opposition towards euthanasia (see next chapter). Finally, they accused the political parties that support euthanasia of having incomprehensible legislative haste and of lacking political legitimacy because most of them did not mention euthanasia in their electoral manifestos. Finally, some of them claimed that a referendum about euthanasia should be held.

Political Party	Position	Main arguments presented
PEV	Positive	– Defending the principle of human dignity; – Palliative care is not similar to euthanasia; – Respect for the objection of health professionals; – Need to avoid pain and suffering; – Constitutional Court did not declare euthanasia, as a whole, to be against the Constitution; – Significant changes were made to accommodate the criticisms/ doubts from the President and the Constitutional Court.
BE	Positive	– Respect for the rights to individual autonomy and freedom of choice; – Defending the principle of human dignity; – Respect for the right to death; – The success of the laws of voluntary termination of pregnancy and drug addiction; – Significant changes were made to accommodate the criticisms/ doubts from the President and the Constitutional Court; – Constitutional Court did not declare euthanasia, as a whole, to be against the Constitution; – Against holding a referendum.
PS	Mostly positive	– The right to life provided in the Constitution is not a duty; – Defending the principle of human dignity; – Respect for the rights to individual autonomy and freedom of choice; – Need to avoid pain and suffering; – The success of the laws of voluntary termination of pregnancy and drug addiction, with no 'slippery slope'; – Significant changes were made to accommodate the criticisms/ doubts from the President and the Constitutional Court; – Constitutional Court did not declare euthanasia, as a whole, to be against the Constitution; – Against holding a referendum.
PAN	Positive	– Defending the principle of human dignity; – Palliative care is not the only solution when suffering is not treatable; – Respect for the right to death; – Respect for the rights to individual autonomy and freedom of choice; – The defence of freedom of conscience and religion is granted; – Significant changes were made to accommodate the criticisms/ doubts from the President and the Constitutional Court; – Constitutional Court did not declare euthanasia, as a whole, to be against the Constitution; – Against holding a referendum.
IL	Positive	– Defending the principle of human dignity; – Respect for the rights to individual autonomy and freedom of choice; – Significant changes were made to accommodate the criticisms/ doubts from the President and the Constitutional Court; – Constitutional Court did not declare euthanasia, as a whole, to be against the Constitution; – Against holding a referendum.
L	Positive	– Respect for the right to death; – Respect for the rights to individual autonomy and freedom of choice;

Fig. 4.2 Positions and arguments on euthanasia from all the political parties throughout the six rounds

		– Defending the principle of human dignity; – Against holding a referendum.
PCP	Negative	– The National Health Service and supply of palliative care need to be strengthened and freely accessible to all citizens before legalising euthanasia; – The danger of the 'slippery slope'; – If approved, euthanasia will become a 'business'; – Respect for the right to life; – Legislative haste; – Against holding a referendum.
PSD	Mostly negative	– Respect for the right to life; – The National Health Service and supply of palliative care need to be strengthened and freely accessible to all citizens before legalising euthanasia; – Lack of political legitimacy because most parties did not mention euthanasia in their electoral manifestos; – The role of the State is to protect the lives of citizens; – The danger of the 'slippery slope'; – Several external entities issued a negative opinion about euthanasia; – The practice of euthanasia will impair doctor-patient relationship; – Legislative haste; – The issue should be decided with a referendum (in the latter rounds). *Vs.* – Need to avoid pain and suffering; – Respect for the rights to individual autonomy and freedom of choice; – Defending the principle of human dignity; – The success of the laws of voluntary termination of pregnancy and drug addiction, with no 'slippery slope';
CDS-PP	Negative	– Respect for the right to life; – The role of the State is to protect the lives of citizens; – The danger of the 'slippery slope'; – The practice of euthanasia will impair doctor-patient relationship; – The National Health Service and supply of palliative care need to be strengthened and freely accessible to all citizens before legalising euthanasia; – Lack of political legitimacy because most parties did not mention euthanasia in their electoral manifestos; – Several external entities issued a negative opinion about euthanasia; – Legislative haste; – The issue should be decided with a referendum.
CH	Negative	– The National Health Service and supply of palliative care need to be strengthened and freely accessible to all citizens before legalising euthanasia; – The danger of the 'slippery slope'; – Several external entities issued a negative opinion about euthanasia; – Respect for the right to life; – Constitutional Court declared the bills as unconstitutional; – Legislative haste; – The issue should be decided with a referendum.

Fig. 4.2   (continued)

NOTES

1. All details regarding this debate are available at http://debates.parlamento. pt/catalogo/r3/dar/01/13/03/090/2018-05-30/2? pgs=2-36&org=PLC
2. All details regarding this debate are available at https://debates.parlamento. pt/catalogo/r3/dar/01/14/01/032/2020-02-21/3? pgs=3-40&org=PLC
3. All details regarding this debate are available at
   https://debates.parlamento.pt/catalogo/r3/dar/01/14/03/01 9/2021-11-04/4
4. All details regarding this debate are available at https://debates.parlamento. pt/catalogo/r3/dar/01/15/01/023/2022-06-09/28?pgs=3-28&org=P LC&plcdf=true
5. All details regarding this debate are available at https://canal.parlamento. pt/?cid=6802&title=reuniao-plenaria

The (Un)used Consultation Processes: Positions and Arguments from Organizational Experts

Abstract This chapter analyses the positions and arguments expressed by thirteen organizational experts which participated in the parliamentary debate through in-person hearings, written opinions (*pareceres*) following a request from the Committee on Constitutional Affairs, Rights, Freedoms and Guarantees, or even through unrequested written contributions.

The analysis revealed that most of these actors were against euthanasia, namely healthcare-related actors, such as the National Council of Ethics for Life Sciences, the Order of Physicians, the Order of Nurses, the Director General of Health, the Portuguese Association of Palliative Care, the Portuguese Association of Bioethics, the Centre of Studies on Bioethics, as well as the Order of Lawyers. On the other hand, the remaining five organizational experts expressed a neutral position, namely the Order of Portuguese Psychologists, the High Council of the Judiciary, the High Council of the Public Prosecution Service, the Portuguese Association of Insurance Providers and the Specialty Council on Clinical and Health Psychology from the Order of Psychologists.

© The Author(s), under exclusive license to Springer Nature Switzerland AG 2023
I. S. Almeida, L. F. Mota, *Politics and Policies in the Debate on Euthanasia*, https://doi.org/10.1007/978-3-031-44588-0_5

The most common arguments were the need to reinforce the supply of palliative care, the incompatibility of medical ethics with euthanasia or the danger of the 'slippery slope'.

Keywords Euthanasia; Experts; Parliamentary hearings; Written opinions; Positioning; Arguments

As shown in Chap. 3, several actors participated in the parliamentary debate[1] on euthanasia, namely through in-person hearings, by issuing written opinions (*pareceres*) following a request from the Parliamentary Committee on Constitutional Affairs, Rights, Freedoms and Guarantees, or even through unrequested written contributions.

These actors can be divided into organizational experts, who are analysed in this chapter, and organizational interest groups, which will be analysed in the following chapter. The organizational experts can, in their turn, be divided into a group of experts who are often asked to produce opinions about morality issues (Sect. 5.1) and those who are not (Sect. 5.2). These opinions were expressed during the first, second and fourth rounds, as the third, fifth and sixth rounds did not include hearings as only amendments to bills were under discussion. Concerning organizational experts, in figure below (Fig. 5.1) it can be seen that 13 organizations expressed their opinion about the topic or about bills.

5.1   The Often-Consulted Organizational Experts

5.1.1   *National Council of Ethics for Life Sciences*

The National Council of Ethics for Life Sciences—an advisory body independent from Parliament with the mission of analysing ethical problems related to biology, medicine or health in general and the life sciences—was one of the first entities to be involved in a hearing, held on June 23, 2016, concerning the **petition 'Right to Die with Dignity'**. The then President of the CNECV, João Lobo Antunes, stated that he would make a personal intervention, as the Council had not yet discussed the petition under consideration.

He considered that <u>palliative and continued care needed reinforcement</u> to meet the needs of end-of-life patients. For this reason, he considered that petitioners were unaware of the numerous possible forms of death

Entity / Person	In person hearings	Rounds	Written opinion	Rounds
National Council of Ethics for Life Sciences	Jun. 23, 2016	R1 (1st petition)	1 (out of 4) requested written opinion	R1
	Apr. 26, 2017	R1 (2nd petition)	1 requested written opinion	R2
			1 requested written opinion	R4
Order of Physicians	Jun. 6, 2016	R1 (1st petition)	5 requested written opinions	R2
	Jul. 16, 2020	R2	4 requested written opinions	R3
Order of Nurses	Jun. 30, 2016	R1 (1st petition)	5 requested written opinions	R2
	Jul. 15, 2020	R2	2 written requested opinions	R4
Order of Portuguese Psychologists	-	-	1 common written opinion	R1
			5 requested written opinions	R2
			4 requested written opinions	R4
Order of Lawyers	May 23, 2017	R1 (2nd petition)	3 requested written opinions	R2
			4 requested written opinions	R4
High Council of Judiciary	-	-	2 written requested opinions	R1
			3 written requested opinions	R2
			1 written requested opinion	R4
High Council of the Public Prosecution Service	-	-	1 written requested opinion	R1
			5 written requested opinions	R2
			2 written requested opinion	R4
General Director of Health	Dec. 6, 2017	R1 (2nd petition)	-	-
Portuguese Association of Palliative Care	Feb. 1, 2018	R1 (2nd petition)	-	
	Jul. 7, 2020	R2		
Portuguese Association of Bioethics	Feb. 1, 2018	R1 (2nd petition)	-	
	Jul. 7, 2020	R2		
Portuguese Association of Insurance Providers	Jul. 15, 2020	R2	1 contribution	R4
Specialty Council on Clinical and Health Psychology - Order of Psychologists	Oct. 6, 2022	R4	-	-
Centre of Studies on Bioethics	-	-	1 written opinion (by their own initiative)	R4

Fig. 5.1 Moments and methods of participation from organizational experts in the debate on euthanasia in Portugal. (Source: adapted version of a summary produced by Parliament services [The document produced by the parliamentary services is available at: https://app.parlamento.pt/webutils/docs/doc.pdf?path=6148523063484d364c793968636d356c6443397a6158526c6379395956 6b786c5a793944454543030764d554e425130524d5279394562324e3162575 6756447397a5357357059326c6864476c3259554e7662576c7a61632324676 4c3245784e544e694d446b5794c544a6d4f5745744e4459794d433034597 a67354c5467334575a6c4e7a55324e6a4d7a5a4335775a47593d&fich=a153b 092-2f9a-4620-8c89-879fe756633d.pdf&Inline=true])

and simplified the depth of the issue, saying the issue needed further debate, as there was a need to discuss not only physical suffering but also spiritual suffering and the challenge it raises for patients.

He also rejected the idea that death has dignity, stating that *"the only thing that has dignity is life"*, which deserves to be preserved until the last moment. In any case, he considered dignity as a useless concept in ethical terms. Finally, he considered that the group of countries where euthanasia is legal (such as the Netherlands, Belgium and Luxembourg) are countries with a very similar origin and cannot be compared to Portugal. In this context, he demonstrated his concern about the danger of the 'slippery slope'. At the end of his hearing, he recalled the existence of the living will and the low number of citizens who had already filled in the form.

In a **second hearing** held with the National Council of Ethics for Life Sciences on April 26, 2017, and under the assessment of the **petition by the movement 'All Life Has Dignity'**, the new President of the CNECV, Jorge Soares, said it was necessary to broaden the reflection on the ethical issues in question, since the members of the Council themselves, given the complexity of the theme, have difficulty in taking a position on the issues.

He started by considering that the reinforcement of palliative care and home palliative care should be addressed in the first place. He also added that the question raised for the CNECV is medical ethics during the process of death, mainly due to the irreversibility of a medical error during the process.

Besides the in-person hearings, the CNECV also participated with different written opinions.

Regarding the **first rounds** of discussion, the CNECV only issued an opinion about the bill from PAN, as it considered that it made no sense to issue opinions about the bills from the PS, PEV and BE when the time of voting would happen in the very near future. Regarding the **bill from PAN**, the CNECV considered, first of all, that the legalization of euthanasia would *"(...) open a gap of relevant ethical and social significance by the asymmetry of the conditions available and the inequities in access to health-care"*. It also considered that the political party's bill seemed to put the respect for death requests and respect for decisions to refuse, abstain or suspend therapeutics on the same level.

The CNECV also rejected the argument related to autonomy, since the final decision will always be taken by physicians. It added that the State should be concerned with ensuring direct access to health-care for end-of-life patients, rather than putting the legalization and regulation of requests for death at the forefront.

Regarding the **second-round** bills submitted by the BE, PAN, the PS and the PEV, the CNECV issued a common written opinion, structuring its reasoning in the following nine arguments:

1. *the lack of studies that support the bills;*

2. *taking care of the people who ask to die should entail relational and integrative answers, which target different dimensions of suffering;*

3. *the State must not assume that euthanasia is a viable answer to deal with suffering when there is a lack of those relational and integrative answers;*

4. *the request to die should be firstly interpreted as a 'cry for help' with several complex meanings, such as "fear, loss of control, loneliness, the feeling of 'burden', or unbearable physical pain";*

5. *there is no knowledge about how many professionals will be available to participate in the necessary procedures, which are currently excluded from the medical praxis;*

6. *the figure of the conscientious objector should not be invoked to acts that are not part of the profession;*

7. *the relationship that the multiple stakeholders (namely physicians, nurses, pharmacists) will develop with the health system and the National Health Service is vague;*

8. *the organizational and financial burdens that the implementation of the bill will imply are unknown;*

9. *the State's responsibility in providing euthanasia procedures without promoting "discrimination of any nature (economic, social, ethnic or geographic) is undetermined".*

Finally, in the **fourth round**, the CNECV decided to also analyse several bills (BE; PS and PAN) at the same time, in a written opinion issued on June 9, 2022. In this joint opinion, the CNECV started by presenting a summary of all its previous opinions and a brief presentation of the new bills. In addition, the opinion presented an ethical reflection on the concepts of "serious and incurable disease" (mentioned in the bills of the BE and the PS) and of "serious or incurable disease" (mentioned by the bills of PAN), which they consider as an inevitable fallacy and redundancy and that the truth is that these new concepts represent a change to respond to the limitation of the presidential veto. In this sense, they added an ethical reflection to the concept of suffering and respect for the patient's autonomy, stating that the existence of free will cannot be guaranteed when access to palliative care is limited. They also made comments on the composition of the committee monitoring the procedure, considering it as disproportionate to include only one physician but two jurists.

At the end of this opinion, the CNECV stated was against these new bills.

5.1.2   Order of Physicians

The Order of Physicians also participated in the debate through two in-person hearings and several written opinions.

The first of these hearings was held on June 30, 2016, **regarding the first petition**,[2] in which the Order of Physicians was represented by its then leader José Manuel Silva, who considered that all lives have dignity and are worthy. Regarding the argument concerning the "total elimination of suffering", he stated that this is a concept that cannot be placed on this level, since *"there is no life without suffering"*. As concerns the concept of 'option', he stated that an actual option would imply that patients would have access to all options, which is not the case regarding palliative care. In this regard, he mentioned the French example of the request of terminal sedation without deliberate induction of death, which is something that should be distinguished from palliative care and enabled to achieve the goal of the petition: *"death with dignity and without suffering"*.

Finally, he considered that *"ignoring the experience of other countries and the so-called slippery slope"* was worrying.

As regards the **written opinions**, it is important to acknowledge, first of all, that the Order of Physicians decided not to issue any opinion regarding the **first-round bills**.

Regarding the **second-round bills**, the Order of Physicians issued separate written opinions about the five bills (BE, PAN, PS, PEV and IL). Analysis of these opinions reveals nevertheless that they are very similar. In these opinions, the Order of Physicians started by mentioning the Portuguese legislation which deals directly with the issues of the final phase of life—Law No. 25/2012, which defines the living will and the Basic Law of Palliative Care. After that initial consideration, the Order continued by reflecting on the arguments used by the political parties to justify such bills, stating they disagreed with the possible link between euthanasia and the constitutional principles of inviolability of the right to life and inviolability of the right to physical and moral integrity.

In addition, they stated that euthanasia and assisted suicide are <u>not medical care practices and are outside the principles of medicine</u>. At the end of the opinions, they declared that *"based on the arguments set out and the invocations made, euthanasia and assisted suicide, under the name «early death», may NOT take place in medical practice following medical law and ethics and medical ethics and deontology"*.

Finally, the Order of Physicians also issued written opinions regarding the **fourth-round bills**. Again, it issued separate opinions but very similar opinions. In these opinions, the Order started by doing a historical review of the discussion of euthanasia in Portugal and the previous opinions expressed by them. Then, they mentioned that the changes introduced in the bills did not change the deontological and ethical fundaments of the proposed practices of euthanasia.

In addition, they stressed the <u>importance of assessing the patient's mental state and the subjective character of the personalization of the diagnosis</u>. At the end of their opinion, they emphasized that such practices <u>do not fit the practice of medicine</u> and that *"euthanasia and assisted suicide can be decriminalized and authorized by law, but do not belong to medicine, they are not configured as medical acts (...) [and that] the Order of Physicians will always continue to strive for the defence of its ethical and deontological values"*.

Besides these opinions, they also presented an older written position, which, after 26 points of arguments, stated the Order would *"refuse to appoint physician(s) to any committee that the legislation provides for and/or to carry out any type of action resulting in a direct or indirect collaboration and/or participation of the Order of Physicians in preparatory procedures and/or the execution of an actor of 'hastening of death on request' or 'medically assisted death'"*.

5.1.3 Order of Nurses

The Order of Nurses also participated in the debates through in-person hearings and written opinions.

The **first of these hearings** was held on June 30, 2016, with its president, Ana Rita Cavaco, who started her intervention by declaring that *"assisted death is all that occurs today in our health system (...) [since] no one dies alone"*. She also stated that the Order of Nurses should recognize the Portuguese population's understanding of what is being discussed, the terms and designations used and its concepts. And, in addition, she ensured that <u>professionals who are against the practice of euthanasia have their position protected</u>. She also stated that health has no undisputable borders, which means that <u>medical acts change over time</u>.

In a **second hearing** held on July 15, 2020, during the **second round of discussion**, this Order was represented by Ana Rita Cavaco (President) and Luís Barreira (vice president), who declared that their intervention would not cover their position on the topic but would mostly be focused on the content presented by each bill. They added that none of the bills had changed their text after the recommendations of the Order of Nurses. At the end of their speech, they stated that they were available to clarify and assist the committee responsible for discussing euthanasia concerning their suggestions.

Concerning the **first-round bills**, the Order of Nurses started by issuing an opinion about the bill of PAN, in which they stated that nurses consider that, before legalizing euthanasia, it is essential to <u>ensure that the Palliative and Continuing Care Network is competent, effective, efficient and enables immediate access</u>. In this regard, the Order stated that, as this network does not function properly, it cannot be guaranteed that the decision for euthanasia is completely free and conscious. They also mentioned that the <u>bill does not properly take into consideration all health professionals but only physicians</u>.

About the bill from the BE, that written opinion stressed, in addition to the arguments referred to for PAN's bill, the <u>necessary revision of the freedom of conscience, religion and worship of professionals</u>, the wording of which was not acceptable.

Regarding the **second-round bills**, the Order of Nurses issued separate opinions regarding all bills (BE, PAN, PS, PEV, and IL) but they are very similar. In these opinions, they once again stressed the need to first <u>ensure a network of effective and competent palliative and continuous care</u>. They also added that the bills remain very simplistic, in the sense that it <u>centralizes the process on a single health professional, a physician, ignoring the role of nurses and other professionals</u>. Nevertheless, it acknowledged the amendments to the Evaluation Committee, which is in line with what is advocated by the Order concerning the relevance of multidisciplinary involvement and the specific participation of some health areas. At the end of all the opinions, the Order stated that they were not in favour of those bills.

As for the **fourth-round bills**, the Order of Nurses maintained the issues and concerns expressed before. They affirmed that the bills had introduced changes, but that there was no substantial difference. They stated that nursing aims to provide healthcare to human beings throughout their lives and that nurses assume the duty to care for the person, <u>defending human life in all circumstances</u>. The Order stated that the <u>lack of provision and access to palliative care</u> and other appropriate care for the patient's situation constitutes an obstacle to the discussion of the bills and that the State itself was deviating from the fulfilment of its obligation to ensure such care and access. At the end of these opinions, they stated their negative opinion on the bills.

5.1.4   Order of Portuguese Psychologists

Unlike the previous actors, the Order of the Portuguese Psychologists only participated in the debate through written opinions.

Concerning the **first round of bills**, the Order only expressed its common opinion about the four bills based on a literature review of several scientific documents. They started by stating that the <u>role of psychologists should be essential to analyse the mental state of patients since requests for early death would be 4.1% higher in depressive patients</u>. Moreover, the Order also warned that <u>people tend to change their minds if they have access to adequate palliative care</u>. They also stated that, once the legislation is approved, *"it is necessary to ensure that patients are accompanied by a multidisciplinary team of which psychologists are part, responsible for evaluating the psychological processes and competencies of individuals to make decisions and give informed consent, as well as providing psychological support to individuals, family members, and health professionals"*.

Regarding the **second round of discussion**, the Order of Psychologists issued its opinions on the projects of the PS, the PEV and the IL, whose contributions it considered relevant for the reflection on the role and importance of psychologists in this issue. They stressed that the Order does not assume a position in favour of or against euthanasia but suggested ways in which psychologists are useful and should be involved in the discussion about the end of life. In this sense, they considered <u>their role to assess the patients' competence and functional capacity to make decisions as essential</u>, as well as supporting patients, family members and health professionals throughout the process. They also believed that the assessment of the "current, serious, free and enlightened will" should be carried out by a psychologist, adding that the <u>opinion of the responsible physician should be accompanied by an opinion of a psychologist</u>, and therefore the presence of a psychologist in the committee responsible should be mandatory. At the end of all opinions, the Order declared that *"(...) the decision on the bills (...) should include the psychological aspects and factors that influence the process (...), as well as the essential role that psychologists and psychological intervention can play in this context"*.

Finally, in the **fourth round of discussion**, the Order of Psychologists issued its opinion on the bills of the BE, PS and PAN, in which it considered that the <u>presence and participation of psychologists in the evaluation committee should be ensured during the procedure</u>. Moreover, they considered to it necessary to ensure that all patients who request assisted death are advised for clinical consultation at the very beginning of the process and that psychological support is offered to other parties such as family members and health professionals. In addition, they recommended that <u>minimum deadlines be defined between the request and the performance of the act</u>.

5.1.5   Order of Lawyers

The **Order of Lawyers**, on May 23, 2017, represented by Isabel Cunha Gil and Pedro Cabeça, members of the General Council, declared on the **petition 'All Life has Dignity'** that *"(...) the Order does not speak, nor will it speak, to the extent that it is a matter and a strictly political, philosophical or ethical option, (...) to be taken by the legislator (...) and from the intimate and personal forum"*. Finally, they considered that, if any legislative measure is adopted, palliative care and the choice of euthanasia should not be confused and should always coexist.

In the **first round of discussion**, the opinion of the Order of Lawyers on the bill of PAN stated that a physician who accesses an application for early death will be practising "active voluntary euthanasia" and that this practice is mentioned as being criminal in the Penal Code as "murder at the request of the victim" (article 134). The Order, therefore, considers that this situation incurs a "crime of incitement or aid to suicide", provided for in article 135. In addition, it exposed several problems in the bill: the relationship between the subject and life itself; the relationship with the third party that assists the patient; the used legal definition of dignified or undignified life; the definition of freedom and individual self-determination; the legal definition of 'unbearable pain'; and the validation of consent itself.

These changes and remarks were also mentioned in their opinion about the bills from the BE, the PS and the PEV.

In the **second round**, the Order of Lawyers did not produce a written opinion concerning the bills from the BE, PAN, the PS or the PEV. Regarding the first two bills, they only produced an information note, which explained that they would were doing so because these new bills were quite similar to those from the first round, which had already been analysed.

The only bill analysed by the Order in the second round was therefore the one from the IL. In this written opinion the Order started by stating that the defence of citizens' rights, freedoms and guarantees constitutes a significant attribution of the Order. In this regard, they claimed that they could not forget the fact that their Order protects human life from conception to natural death, justifying that *"life is inviolable and that no one can dispose of their life, as no one can alienate their freedom or respect for themselves"*. This leads to the question of the legal plan and the irreducibility of the personality rights of the human being. They also stated, in response to the argument of freedom and individual choice, that personality rights are renounceable, but that they do not believe that life, an inviolable right, constitutes an indispensable and inviolable right, as well as other personality rights. Concerning palliative and continuous care, they asked whether they were properly provided, freely accessible to all citizens, and given the lack of investment in these domains, whether the will of the patient could still be considered as current, free, serious and enlightened. They also considered it impossible to ignore the fact that no technical advice in favour of the decriminalization of euthanasia is known. At the end of their opinion, the Order declared that the bills were not in line with the legal and constitutional frameworks.

Regarding the **fourth-round bills**, the Order of Lawyers did not issue a written opinion, but only an information note in which they considered that none of the bills was significantly different from the first and second-rounds bills, and added that they had not changed their mind, which meant there was no justification for amending their previous opinions.

5.1.6   High Council of the Judiciary

Regarding the **first-round bills**, the High Council of the Judiciary issued written opinions about the bills from PAN and BE, in which they stated that changes to the Criminal Code are necessary and made a brief reference to the issue of human rights and European comparative law, such as the right to life, the right to respect for private and family life, the right to avoid undignified death, the right of individual self-determination and others. However, in their conclusion on the bills, they stated that the Council does not have attributions and competencies concerning the ethical-philosophical and political-legislative considerations under discussion.

As concerns the **second-round bills**, the High Council of the Judiciary issued opinions only about those from the BE, the PS and the PEV, which were very similar, while for PAN's and IL's bills, the Council declared only that it had already adopted its previous opinions and there was nothing to add to the new bills. In the opinions, they addressed the sensitivity of the topic and the intellectual gravity that it requires, yet they considered that it was not up to the Council to *"(...) assess, nor take sides in such positions, especially since it is a matter of eminently political and philosophical choice, of an individual and social ethical nature"*. They considered that the Council had already ruled on the issue in previous opinions, making reference to the national constitutional and legal framework and to the decisions of the European Court of Human Rights.

Concerning the **fourth round**, the High Council of the Judiciary only issued a written opinion about the bill of the BE, excusing itself from doing the same regarding the bills from the PS and PAN considering the similitude of the bills. In this opinion, the Council recalled that the issue is sensitive and intellectual gravity is necessary in approaching it and asserted that, although two different opinions may be considered on fundamental values and their defence, it is not for the Council to assess or take sides, assuming that it is a political and philosophical matter of an individual and social ethical nature. Regarding the concrete assessment of the bill, they assumed that the doubts about the concepts, raised by the President in his veto, had been answered.

5.1.7 High Council of the Public Prosecution Service

Regarding the **first-round bills**, the High Council of the Public Prosecution Service only issued a written opinion about the bill from the PAN. In this opinion, the Council used four main arguments, of which the amendment to articles 134 and 135 of the Penal Code can be highlighted, admitting the adoption of a point justifying that *"the act is not punishable if it has been practised by a physician following the law governing access to medically assisted death"*. For this reason, they considered it essential that the law also ensured the protection of nursing assistants, especially that they no punished, since their non-inclusion may leave the conduct of these professionals out of the frameword. Finally, it considered that several articles of the bill should be reviewed, namely regarding the terms used.

Regarding the **second-round bills**, the High Council of the Public Prosecution Service issued opinions about the five bills—that of the BE, the PS PAN, the PEV and the IL—but they are all very similar. In these opinions, they stress some topics to be examined, such as the clarification of concepts that leave some margin of subjectivity, such as the concepts "permanent injury or incurable and fatal disease" and "lasting and unbearable suffering", as well as the amendment of the Penal Code. They added that it would be important for the responsible physician to be a specialist or a physician that knows the patient's clinical history. They also declared that *"the object of the bill constitutes a political-legislative option (...) which summons (...) philosophical, ethical, or even religious convictions, on which we must not express an opinion"*.

Concerning the **fourth-round bills**, the High Council of the Public Prosecution Service issued an opinion about the bill from the BE and a joint opinion about the bill from the PS and PAN, which were very similar. In these opinions, the Council acknowledged that the small changes introduced in the new bills aimed to overcome the objections of the November 2021 political veto. After a brief presentation about the declaration of unconstitutionality by the Constitutional Court and the presidential veto and a summary of the bills, they stated that the Constitutional Court issued a clear pronunciation, even if not with a unanimous majority, as to the link and constitutional reconciliation between euthanasia and the right to life and other fundamental rights. At the end of their opinions, they asserted that those bills established a proper possible revision of the Penal Code.

5.2   THE LESS COMMONLY CONSULTED ORGANIZATIONAL EXPERTS

Besides the mentioned entities which are often asked to provide written opinions about various topics under discussion in Parliament, other less commonly consulted organizational experts were also involved through in-person hearings to which they were invited.

5.2.1   *Director General of Health*

One of the people consulted was the Director General of Health, Graça Freitas, whose hearing was held on December 6, 2017, during an early stage of the debate of the **first round and concerning the second petition**. She began her intervention with the topic of palliative care, asserting the following:

> *(…) the State should be more responsible for the quality of life of their citizens, particularly during their last period of life. (…) [and that] the dominant culture of society has considered the cure of the disease as a priority goal of its health services, and the incurability and the inevitable reality of death are often neglected in the daily practice of care.*

She proceeded by stating that the role of health professionals is to accompany patients at the end of their life and to help them with their suffering and symptoms. For this reason, she believed health professionals should acquire specific skills to help patients and alleviate their concerns. Furthermore, she stated that the response to a patient cannot be euthanasia, assisted suicide or excessive interventionism, as the answer will always be to help the patient through palliative care on the principle of equity. And she questioned whether euthanasia will still be necessary if a point is reached in which palliative care is widespread, with full professional care. To finish her speech, she stated that *"(…) when there is global, good, equitable palliative care, (…) and in the face of physical, psychic, affective comfort"*, the need for euthanasia can then be discussed.

5.2.2   *Portuguese Association of Palliative Care*

The Portuguese Association of Palliative Care, represented by its president, Duarte Soares, participated in a hearing held on February 1, 2018, during the **first round of discussion**. He began his speech by saying that, in addition to being a signatory to the petition of the Portuguese Federation of Life (against euthanasia), the association itself is following its parameters. He proceeded by mentioning that *"(...) the CRP determines that human life is inviolable"*. He then asked all those involved in the debate for wisdom since this is a clinical, social and value issue and not a political issue, which means the *"mediatization of the pros and cons"* should be avoided.

At the end of his speech, he asked for caution over the terms used in the bills and the way they are defined, since what bills establish *"(...) is not a right to die, or of having an assisted death, it is a right to be killed by another person (...) It is not, therefore, a dignified death, it is a murder on request and that (...) does not define the dignity of death"*. He, therefore, considered that all end-of-life situations are not covered, so existential suffering will not be clearly defined.

This association was also present in a second hearing, held on July 7, 2020, during the **second round of discussion**, represented by Teresa Sarmento, who declared there are new challenges and a greater number of patients who need palliative care, raising a question about how to act on this suffering.

She also mentioned that the association seeks to *"promote palliative care (...) [and therefore] to improve the quality of life of patients and their families (...) by preventing and relieving their suffering"*. After this statement, she proceeded by saying that the existent palliative care in Portugal is insufficient, as less than 20% of patients have access to it. From this perspective, she asked if *"(...) in the true promotion of quality of life, can we speak of a true freedom of choice, when we offer the possibility to end your life early if we do not give you the possibility to live your life with dignity because we cannot afford it?"*

At the end of her hearing, she recalled that euthanasia eliminates the person who suffers and does not eliminate suffering and recalled the importance of palliative care and the need to train more people to work in this field.

5.2.3   Portuguese Association of Bioethics

As with the previous entity, the Portuguese Association of Bioethics was also involved in a hearing held on February 1, 2018, during the **first round**, in which it was represented by its president Rui Nunes, who began his intervention by mentioning that *"(...) the human person is a subject with undisputed dignity and rights, and it is worth affirming this in society"*.

He also stated that public authorities, and Parliament in particular, should look at end-of-life issues and that, for that reason, palliative care with quality should be ensured, with properly qualified health professionals. He added that a well-designed and structured palliative care network will enable money to be saved for taxpayers.

Regarding euthanasia and assisted suicide, he said that *"it is a matter of great complexity (...) and that the Constitution would have to be refined to contemplate any legislative change in this matter"*. He, therefore, considered that a national referendum should be held on active and voluntary euthanasia. Finally, he questioned who could administer it, but this is another debate to be held in the event of approval of the decriminalization of euthanasia.

The association was also present in a second hearing, held on July 7, 2020, during the **second round of discussion**, represented again by its president Rui Nunes, who declared that a death certificate should always contain the near cause of death and not the underlying disease, for example, a quadriplegic patient who asks for early death, cannot have tetraplegia as a cause of death.

In addition, he considered that there is a duty to inform patients about the existing instrument of the 'living will'. He took this opportunity to stress that the bill from the BE project—which stipulated that the procedure of euthanasia could continue even if the patient was unconscious if the living will mentions this possibility—required an amendment of the living will law itself, which in article 5 prohibits the practice of euthanasia. In this context, he questioned why the process would continue only if the living will specifies it and not if the patient is represented by a "health care attorney".

At the end of his hearing, he stated that he did not understand why a patient could ask a health professional to administer the lethal drug but could not ask a healthcare attorney or a close friend or relative to do so.

5.2.4   *Portuguese Association of Insurance Providers*

During the **second round of discussion**, the Portuguese Association of Insurers were also involved in a hearing, which was held on July 15, 2020, with Alexandra Queiroz (director-general) and José Galamba de Oliveira (president) as its representatives.

This hearing assumed particular importance since, unlike what had happened in the first round, the second-round bills were dealing with the topic of insurance. In this sense, they proposed several amendments to the text of the corresponding article in the bills of the BE, the PS and PAN—for example, where it was written *"the health professionals who participate (...) in the clinical procedure of hastened death"*, the wording should be *"the people who participate"* so that the bills are more comprehensive and encompassing. Another suggested amendment concerned the death certificate, in which they considered that the reported cause of death should be the pathology that led patients to ask for euthanasia rather than *"early death"*, so that it becomes clear that the relevant cause of death for contractual purposes is not the medically assisted death procedure itself, but the pathology that was at the origin of it. At the end, they added a new point to the article, which stated that insurance companies should have *"if companies are insured, they have access to the necessary information to comply with the provisions of this article in accordance with the conditions established in regulations to be approved by the Government"*, to speed up the process of paying compensation and avoiding unnecessary litigation.

5.2.5   Speciality Council on Clinical and Health Psychology of the Order of Portuguese Psychologists

The Specialty Council on Clinical and Health Psychology of the Order of Portuguese Psychologists was also present in a hearing in Parliament, namely on October 6, 2022, during the fourth round of discussion, represented by the president, Miguel Ricou, who stressed they had three distinct concerns related to issued of mental health of competence.

He mentioned they were mainly concerned *"(...) about a dimension (...) which has to do with (...) the ideation of death in euthanasia (...) and the ideation of suicide"*. He stressed that, even within suicide, there is a universe of dimensions, from the will to die to the attempt, and the same would happen in the universe of assisted death, since the desire to die does not give patients the direct access to the procedure. This means all these people need to be followed by specialists, such as psychologists, since the will of the individual can change over time and, in these cases, errors are not admitted, considering this as the second dimension. The third dimension concerned developments that may or may not exist in these cases, and therefore he considered that there should be a minimum period between the application and the initiation of the procedure of the respective act, which is different for people with a terminal illness and people with an incurable disease since they entail different adaptation processes.

At the end of their hearing, he also expressed concern about the growth of cases in countries where active euthanasia is legal, giving examples of euthanasia in minors.

5.2.6   Centre of Studies on Bioethics

During the **fourth round of discussion**, and after the approval of the bills in general terms, the Centre of Studies on Bioethics, a multi-institutional research unit, issued a written pronouncement on September 15, 2022. They explained that their pronouncement was necessary due to the most difficult moment that Portugal has experienced in the last one hundred years, with the approval of euthanasia in Parliament and, therefore, the institution was making its position public on two main points about euthanasia.

On the topic of euthanasia, they reaffirmed the <u>intrinsic dignity of the human being</u> and that *"this dignity is not dependent on any other circumstance and does not change or undergo any change with age, illness or proximity to death"*. Thus, they considered that arguments centred on autonomy and self-determination to justify euthanasia are a fallacy since life is the support of the exercise of freedom.

They also stated that the role of health professionals is the care of patients, with their prevention, diagnosis, treatment and follow-up and with the best means available. They explained, therefore, that *"killing the sick, with the argument of eliminating suffering, is a betrayal of all the effort that health professionals make daily"*. Under these circumstances, they argued that <u>euthanasia is not a medical act</u> and that medical ethics can never involve killing a patient under any circumstances.

They recalled that the <u>Portuguese Constitution stated that human life is inviolable</u> and that the approval of such a law violates the fundamental law, opening a gap in society and constituting a civilizational setback. In this perspective, they considered it urgent to reaffirm the intrinsic value of life, humanize death and the process of death, reflect on the issue of the meaning of life, not excluding disease and suffering, and develop competencies in health professionals regarding the follow-up of patients in terminal phases.

On the timing of the approval, the Centre of Studies on Bioethics argued that the country's efforts should be towards the prevention of the pandemic and in the care to be provided to patients and, therefore, they did not understand why the issue is on the agenda and considered it scandalous that it had occurred. They ended their pronouncement by stating that *"a person with an incurable disease is never an incurable person"*.

Notes

1. All the hearings, as well as requested and non-requested written opinions are available through the links for the bills and petitions provided in Chap. 3. The document whose link is provided in the following footnote also contains direct links.
2. The Order of Physicians was also heard regarding the second petition, but no record of this hearing was found.

Does My Voice Also Count?: Positions and Arguments by Interest Groups in the Parliamentary Debates

Abstract In this chapter, we analyse the positions and arguments expressed by twenty interest groups which participated in the parliamentary debate through in-person hearings and unrequested written contributions. These interest groups may be divided into three categories: three petitioners; nine interest groups dedicated to causes related to euthanasia; and eight religious interest groups.

It was possible to conclude that eighteen of these interest groups expressed their opinion against the legalization of euthanasia, while only the civic movement 'Right to Die with Dignity', which organized the first petition, was in favour of legalizing euthanasia, and the representative of the European Platform "Wish to Die" expressed a neutral position.

The most common arguments against euthanasia were the need to reinforce the supply of palliative care, the inviolability of life or the danger of the 'slippery slope'. Religious interest groups also mentioned the sanctity of life. The arguments in favour of euthanasia were the autonomy of patients, the mercy in ending unbearable suffering, as well as the protection that the proposed bills provide against abuses or the 'slippery slope'.

Keywords Euthanasia; Interest groups; Parliamentary hearings; Positioning; Arguments

I. S. Almeida, L. F. Mota, *Politics and Policies in the Debate on Euthanasia*, https://doi.org/10.1007/978-3-031-44588-0_6

In this chapter, we analyse interventions[1] by what we consider 'organizational interest groups'. This group comprises, first of all, the civic movements and organizations which submitted the petitions that propelled the political and parliamentary debate on euthanasia, which had a huge importance in the first round, but also other 18 civil society actors that participated through in-person hearings and also through unrequested written opinions submitted to Parliament, particularly in the second and fourth rounds of discussion. All the 20 interest groups which participated during the six rounds can be seen below, as well as the moment and method of participation (Fig. 6.1).

6.1   THE PETITIONERS

6.1.1   Civic Movement for the Decriminalization of Assisted Death

One of the entities that most contributed to the political debate on euthanasia was the Civic Movement for the Decriminalization of Assisted Death (the civic movement first named 'Right to Die with Dignity') since it was this civic movement which submitted a petition (Petition 103/XIII/1) to Parliament requestion a discuss of the topic of euthanasia (see Chap. 3, Sect. 3.1.1 of this book).

This civic movement was also present in the committee's hearings twice, the first time when its petition was discussed (June 22, 2016) and the second time when the opposing petition was being discussed in the committee (February 1, 2018).

In the **first hearing**, the civic movement was represented by three people—the physicians Jorge Espírito Santo and Bruno Maia and a teacher, Ana Figueiredo—who declared that the freedom, self-determination and personal choices of citizens were at stake. However, they considered that moral and behavioural issues should not be disregarded during the process.

They also declared that the right to life with dignity is equal to the right to death, as well as the right to individual choice and the possibility of *"leaving in peace and leaving no trauma to any family member"*. They assumed, in this regard, that it is not an easy process and that it should not be done *"to someone who feels uncomfortable or depressed"*, and exceptional criteria should be defined in the legislation to ensure the most appropriate solutions, allowing value to be added to life, and the freedom to make choices in a conscious and informed way. They also ensured that

Entity / Person	In person hearings	Round	Written opinion	Round
Civic Movement for the Decriminalization of Assisted Death	Jun. 22, 2016 Feb. 1, 2018	R1 (1st petition) R1 (2nd petition)	Petition 103/XIII/1	R1
Portuguese Federation for Life	Apr. 19, 2017 Jul. 9, 2020	R1 (2nd petition) R2	Petition 250/XIII Document submitted during the hearing	R1 R2
Movement STOP Euthanasia	Jun. 29, 2017 Jul. 9, 2020	R1 (2nd petition) R2		
National Commission Justice and Peace	Feb. 1, 2018	R1 (2nd petition)	-	
Association of Portuguese Catholic Physicians	Jul. 1, 2020	R2	-	-
Interreligious Working Group on Religion & Health	Jul. 1, 2020	R2	-	-
Portuguese Association of Catholic Jurists	Jul. 8, 2020	R2	-	-
Portuguese Cáritas	Jul. 8, 2020	R2	-	-
Together for Life Association	Jul. 9, 2020	R2	1 document submitted during the hearing	R2
Voiceless Children Movement	Jul. 9, 2020	R2	-	-
First subscribers of the Petition 48/XIV/1 – Referendum on euthanasia	Jul. 15, 2020	R2	Petition 48/XIV/1	R2
Miguel Ricou – Coordinator of the European Platform "Wish to Die"	Jul. 15, 2020	R2	-	-
Civic Movement "Coimbra for Life"	Jul. 15, 2020	R2	-	-
Portuguese Union of Seventh-day Adventists	-	-	1 contribution	R4
Portuguese Buddhist Union	-	-	1 contribution	R4
Portuguese Association of Catholic Psychologists	Sep. 15, 2022	R4	1 written opinion	R4
Hurray-for-Life Association	Sep. 22, 2022	R4	1 written opinion	R4
InFamily Association	Sep. 22, 2022	R4	1 written opinion	R4
Association for the Defence and Support of Life - Aveiro	Sep 29, 2022	R4	1 written opinion	R4

Fig. 6.1 Moments and methods of participation by interest groups in the debate on euthanasia. (Source: adapted version of a summary produced by the Parliament's services [The document produced by the parliamentary services is available at: https://app.parlamento.pt/webutils/docs/doc.pdf?path=6148523063484d364c79396863 6d356c6443397a6158526c63793959566b786c5a793944543030764d554 e425130524d5279394562324e31625756756447397a5357357059326c6864476c 3259554e7662576c7a633246764c3245784e544e6694d446b794c544a6d4f57457 44e4459794d433034597a67354c5467334f575a6c4e7a55324e6a4d7a5a4335775a 47593d&fich=a153b092-2f9a-4620-8c89-879fe756633d.pdf&Inline=true])

euthanasia is not a medical treatment, but it should be a medical act *"and that no physician should be excluded from this discussion"*, and the whole process is based on the doctor-patient relationship strengthened throughout the procedure.

They accepted the fact that the lack of access to palliative care could influence the choice of citizens and that patients could be *"pushed into this process in the face of a more advanced disease"*, rather than going to palliative care in the first place. However, they considered that access to palliative care is a problem that must be addressed by the Portuguese government.

In the **second hearing**, the civic movement was represented by four people—Carlos Salgueiral, Gilberto Couto, Jorge Espírito Santo and Rafael Gonçalves—who declared that this is an essential debate on civilizational issues and human life, arguing that this is an extremely important debate.

They also considered that *"there is no confrontation, no incompatibility, nor any difficulty in reconciling the development of palliative care, (...) which is indispensable (...), with the question that is posed by patients for whom palliative care no longer serves"*. In this context, they considered acts of euthanasia may in no way be considered as 'murder', since this is to attack and belittle patients' lives, acting against their will. Furthermore, they considered that the 'slippery slope' is not scientifically proven in countries where active euthanasia is legal, even if the eventual/reported failures and abuses in these countries should be a point of reflection for our legislation.

At the end, they declared that there is no question regarding the inviolability of human life and autonomy, stating that *"without autonomy, there is no human life"*. In this sense, they stated that the exercise of our full life is done through choices that we make autonomously, and the patient has the right to refuse treatment, even knowing that this will lead to his/her death.

6.1.2   Portuguese Federation for Life

As previously described (see Chap. 3, Sect. 3.1.2), a petition against the decriminalization of euthanasia was also submitted to Parliament in an early stage of this discussion (Petition 250/XIII). This petition was submitted by the Portuguese Federation for Life, which participated in committee meetings twice to present their petition and to express their opinions about the topic.

The first time this entity participated in a **hearing** was on April 19, 2017, and was represented by five people: Maria Isilda Pegado (president of this entity and former MP of the PSD), António M. Pinheiro Torres (vice-president), José Maria Seabra Duque (jurist), Germano de Sousa (physician and former president of the Order of Physicians) and Luís Marques da Costa (director of the Oncology Service of St. Maria Hospital). They declared that their opinion on the topic was based on <u>human dignity</u> and that the <u>State's role should be to protect the least protected citizens</u>, and that all the effects and consequences of the approval of euthanasia, including for friends and family, should be addressed thoroughly.

During their intervention, they also exposed the <u>danger of the 'slippery slope'</u>, presenting concrete data from Belgium and the Netherlands that point to the danger that euthanasia will become an attainable procedure for all citizens. In addition, they stated that this is a procedure that <u>destroys the physician-patient relationship</u> since health professionals are bound to obligations and principles and should respect the lives of their patients and family members. In this regard, they mentioned that, in many cases, the role of a physician is not to heal but to assist their patients.

On the other hand, they mentioned that euthanasia is a <u>form of selfishness</u>, in the sense that is seen as a solution by *"healthy people, who do not want to look at the suffering of others"*. They also considered that witnessing the suffering of others is complex, but that citizens should not be selfish and <u>should use their efforts to strengthen the supply of palliative care</u>. In this regard, they considered that, once the law would be approved, *"(...) the investment in care, ... in para-research to alleviate suffering, will be mitigated because there is a solution that is much easier"*.

Finally, setting out concrete figures on the annual growth of oncological diseases, they considered that <u>investment in palliative care is insufficient in Portugal</u> and that this investment should be increased.

In their **second hearing**, held on July 9, 2020, during the discussion of the **second-round bills**, the representatives from the Portuguese Federation for Life—António Pinheiro Torres and Isilda Pegado—declared that they work for the defence of life and that all bills under-appreciation *"define the authorization or support to kill human beings"*.

They also stated that these bills are <u>against the constitutional right that specifies that *"human life is inviolable"*</u>, which is a very clear one, unlike other more subjective fundamental rights. They thus considered that the approved bills are an attack on this right, as no person should be able to kill another one in our civilization, comparing this situation to the abolition of the death penalty.

In this regard, they stated that the bills are the <u>means to destroy and end our society</u> since assisted suicide or murder on request are now penalized by the Penal Code, even for the most fragile and needy lives. Finally, they considered that the decriminalization of euthanasia is to give the power to some to kill others, under the protection of the law, and to give the State the power to kill the most deprived, thus creating <u>a utilitarian view in society regarding life for economic and even family reasons</u>.

6.1.3   First Subscribers of the Petition 48/XIV/1—Referendum on Euthanasia

On February 19, 2019, petition no. 48/XIV/1 was submitted to Parliament, with had 3938 signatures. The hearing of petitioners was held on July 15, 2020, with Dinis da Silva Freitas, a retired university lecturer in medicine—who declared that this petition had been first sent to the President.

During his speech, the petitioner called for a legislative referendum, saying this would force citizens to reflect on the issue, also asserting his position against the decriminalization of medically assisted death.

He added that the <u>right to human life is an inviolable right</u>, which precedes the right of autonomy and freedom, that <u>life is inviolable</u>, and that <u>there are no lives unworthy of being lived</u>. In this context, he stated that there should not be the right to death, arguing that, with euthanasia, suffering will not be eliminated, but rather the person who suffers.

He expressed <u>concern about the phenomena of the 'slippery slope'</u> in the Netherlands and Belgium, making commonplace what should be an exception, and stated that the autonomy of patients turns out to be a fallacy. He also stated that euthanasia, in addition to <u>not being a medical act</u>, runs the risk of being seen as a response to disease and suffering in this new context where death is trivialized, <u>harming the physician-patient relationship</u>.

At the end of his speech, he criticized the moral and ethical principles of those who advocate the approval of euthanasia, adding that the <u>network of palliative care and continued care needs to be improved</u> and available to all citizens.

6.2 The Interest Groups Focused on Specific Causes

6.2.1 Movement STOP Euthanasia

A few months after the first hearings, the Movement 'STOP Euthanasia' participated in a **hearing**, represented by eight people—António Proa, António Gentil Martins, Vasco Magalhães Ramalho, Alice Menezes, Raquel Abreu, Sofia Guedes, Graça Varão and Thereza Carvalho. During this hearing, they declared that *"(…) all the presidents of the order of physicians consider it essential, for the time being, to clarify for the people and the population what the various terms mean and not to mask the basic ideas with more sympathetic words"*.

They also stated that *"ethically, and for any physician, (…) euthanasia is a violation of the principle of medical ethics"*. For this reason, they understood that euthanasia should be avoided and punishable by law. They also considered that human dignity is associated with life itself and not by *"facilitating the process"*, even in a situation seen as a limit. In this regard, they presented a 'Letter of Health Professionals against the Legalization of Euthanasia', which had 190 signatures.

They expressed concern about the specific cases of countries where euthanasia has been decriminalized, which over the years have extended the number and types of people. Based on this argument, they also presented statistical data from Belgium and the Netherlands. They mentioned that their movement is concerned with people's suffering, particularly of vulnerable people. They have therefore specified that the State must ensure this monitoring, particularly in terms of palliative care, which has a long way to go and that, therefore, they should not be precipitous in the discussion over euthanasia. At the end of their hearing, they expressed concern about the existence of unrecognized dignities, abandoned sufferings and lost of meaning in life.

The movement 'STOP Euthanasia' also participated in a **second hearing**, held on July 9, 2020, during the debate on the **second-round bills**, in which it was represented by Graça Varão and Nelson Brito. During this hearing, they declared that to be in contact with the suffering of others entails questions that, at the outset, have no answer, such as why people suffer. In this sense, they stated that *"(…) no one likes to suffer, (…) Those who agree to live with suffering, usually try to reduce it (…) and do not do so by masochism, nor to receive gratitude from someone else"*.

They also assumed that the pandemic has strengthened the role of health professionals, that is *"(...) to heal and treat, whenever possible, those who suffer"*, making collective learning more evident in facing suffering and in strengthening the sense of community among citizens. In this sense, they considered it important to take care of all people, even those *"(...) who have a permanent injury, incurable disease, or lasting and unbearable suffering"*. Under these circumstances, they considered two paradoxes were at stake: (i) The first concerns physicians, who are always expected to treat, care for and cure the sick, but who, if euthanasia is decriminalized, may become an agent of early death, which <u>may jeopardize the trust in physicians</u>; (ii) by allowing medically assisted death, *"we will be giving a social and political signal of capitulation in the face of suffering and letting suffering command life"*, which may be compared to a scenario which allows all people to use weapons because the State is not able to face violence and crime.

At the end of their speech, they explained that "medically assisted death" is what physicians already try to do and what families and the patients themselves expect from the physicians—that is, assisting the sick individuals. They added that physicians do not have sufficient human and technical resources to provide such aid, considering the <u>poor conditions of the continued care and palliative care networks</u>. Finally, they clarified that, if approved, euthanasia cannot be designated as 'medically assisted' and should not rely on physicians as a professional group.

6.2.2   Portuguese Cáritas

On July 8, 2020, the *Portuguese Cáritas* was represented by Paulo Marques de Magalhães Ramalho and Maria do Céu Patrão Neves in a **hearing about the second-round bills**. During this hearing, they declared they were undoubtedly in favour of life and, therefore, against any form of its interruption, as well as any therapeutic behaviour that aims to prolong life without purpose and at any cost.

They also expressed their disagreement with the bills and concern about the urgency of putting the issue on the agenda and approving it, as well as the <u>absence of legislative initiatives that create a network of palliative care and continued care with free access for all citizens</u>. They assumed that a palliative care network for all citizens would meet many cases covered by the bills under discussion. They stated that the existing national networks are insufficient and that there is no alternative for patients.

. At the end of their intervention, they stated that there are weaknesses in the bills under discussion, from the excessive concentration of decisions in the guiding physician, who does not need to be a specialist in a patient's illness and can even dispense with the opinion of a psychiatrist. They added that the possibility of being able to carry out the procedure anywhere, even at home, raises difficulties in monitoring.

6.2.3   *Together for Life Association*

Another entity participating in the **second round of debates** was the Association Together for Life, which participated in a hearing held on July 9, 2020, represented by Pedro Libano Monteiro and Teresa Melo Ribeiro. They started their intervention by declaring their *"(…) convinced and well-founded opposition to the intended decriminalization and legalization of euthanasia and assisted suicide in Portugal."*

They thus argued that <u>all lives have dignity</u>, which means *"(…) the role of society and the State must be to always assist and care, in all circumstances, and not to kill"*. They added that <u>the State's human, technical and financial resources should be channelled into the provision of health care</u> and not hasten the unnatural death of the patient. In this regard, they argued that *"killing a person or helping them to commit suicide does not end suffering, it ends life, (…) ends the dignity, freedom, and autonomy of the person"*.

They stated that both the National Council of Ethics for Life Sciences and the Order of Physicians expressed several ethical, medical and deontological concerns about euthanasia and assisted suicide, as <u>they are *"not, nor can they be considered, medical acts*</u>, *(…) cannot take place in medical practice"*.

Concerning legal reasons, they stated that the bills under discussion provide for amendments to the Penal Code, which is a necessary precondition so that *"(…) certain uses of death can be legalized and institutionalized at the request of the victim, whether the death is carried out by a health professional, or by the patient, but with the help of another"*. For this reason, they stated that the <u>decriminalization of euthanasia is contrary to the national legal order and the Portuguese Constitution</u>.

Later, they added that it was <u>up to the State, the *"guardian of fundamental human rights"*, to guarantee and defend human life and integrity</u>, in any situation, especially those associated with fragility, disease and suffering. They recalled that the State must be responsible for promoting the provision of primary care, continuous care networks and palliative care networks, thus strengthening the National Health Service.

6.2.4   The Voiceless Children Movement

The 'Voiceless Children's Movement—which advocates for the rights of people with disability and their families—was another entity which participated in a hearing, held on July 9, 2020, during the second round of debates, being represented by Conceição Lourenço and Sara Lourenço.

During their intervention, they declared that they were present with a patient who has already requested euthanasia, stating that patients who request euthanasia not because of their physical suffering related to the disease but due to their situation and the lack of State support. Therefore, they considered that before moving towards a law of decriminalization of euthanasia, we must question whether the State has given sufficient support to these patients—such as palliative care—and only after that they can make a conscious choice.

Giving the example of the IL bill, they stated that the bill assumes it to be physicians who decide whether patients meet the criteria. In this sense, they asked how physicians can be sure about how patients feel since they are not living the same experience as their patients.

At the end of their speech, they said that the lack of State support makes people feel like a burden and therefore end up giving up their lives.

6.2.5   Miguel Ricou—Coordinator of the European Platform 'Wish to Die'

On July 15, 2020, during the **discussion of the second-round bills**, Miguel Ricou, coordinator of the European Platform 'Wish to Die', was present at a hearing, where he began by mentioning that this is a platform of researchers who reflect on the topic of euthanasia.

He presented two dimensions of analysis, the first concerning terminal disease and definitive injury which he considered two situations with many differences. In common, the two dimensions are concerned with mitigating the suffering and pain of the patient, and if the patient can adapt to the circumstances to which the disease obliges. He stressed that there are patients who do not adapt to these circumstances, and that the difficulty of health professionals lies in identifying this.

The second dimension concerned the doubt about decisions. Therefore, it is important for patients to have support, someone to share their doubts and difficulties with, and can discuss the issue. From this perspective, he stressed the importance of psychological intervention, and he claimed that only the bill from the IL mentioned it, albeit briefly.

At the end of the intervention, he considered the role of health professionals to be fundamental for the responsibility of the situation of the death of another, and these professionals should also have access to psychological support. He claimed that in other countries where euthanasia is legal, laws look only at medicine and law, but disregard the role of psychology.

6.2.6 Civic Movement 'Coimbra for Life'

On July 15, 2020, in their hearing during the **second round of discussion,** the civic movement 'Coimbra for Life', represented by João Barbosa de Melo and Maria Teixeira Pimparel, declared that everyone shares meanings of respect for others and respect for those who suffer, and each tries to do their best to reduce the atrocious and senseless suffering of the other, respecting each person's freedom.

They therefore questioned whether we will have the right to end our own lives in the way and in time we see most fit, and in the case of a positive response, if the State and society should be obliged to comply with or help to fulfil this self-determination of the will to die. They recognized, in this sense, that the State and the community have to actively work so that no citizen should give up and want to die.

They also wondered whether Portugal will be able not to make the same mistakes as countries where active euthanasia is legal, mentioning the 'slippery slope'.

At the end of their speech, they accused the country of not investing in palliative care and that the National Health Service does not have sufficient means.

6.2.7 Hurray-for-Life Association

On September 22, 2022, the Hurray-for-Life Association participated in a hearing and delivered a written opinion in the context of the appreciation of the fourth-round bills.

In several points of reflection, they asserted that the bills attribute the concept of 'suffering of great intensity' exclusively to the patient, considering that this perception oscillates and is not resolved with the finitude of life. From this perspective, they assumed that psychological evaluation and medical follow-up are crucial in finding the true origin of the patient's suffering.

Regarding the bills, they considered it important to guarantee access to palliative care, which does not happen in Portugal, not even in the private sector.

At the end of their intervention, they reflected on the absence of follow-up of family members and the stages of mourning.

6.2.8   *InFamily Association*

Like the previous association, the Association *InFamily* also participated in a hearing and submitted a written pronouncement on September 22, 2022.

They started by saying the association's main objective is to promote the defence of life, as well as the protection and promotion of human rights, among others.

They presented their letter of principles, in which they asserted that human life is inviolable and that dignity must be recognized from conception to natural death. Therefore, they believe the State and its institutions, civil society and every citizen have an obligation to promote a culture of respect and to protect human life appropriately.

For the association, the bills under-appreciation attack life, because they constitute an administrative procedure for the sole purpose of executing the death of others, and they attack social responsibility because they entail an unacceptable lack of accountability of the State towards the lives of people, and, they attack the family because they do not include the duty for mandatory dialogue with the relatives of the patient.

At the end of their pronouncement, they stated that the bills on euthanasia that were under discussion did not provide for the participation, involvement and mandatory information of the patient's relatives during the procedure.

6.2.9   *Association for the Defence and Support of Life—Aveiro*

On September 29, 2022, the Association for the Defence and Support of Life—Aveiro participated in a hearing represented by Belmiro Fernandes Pereira, Teresa Soares Correia and Luís Manuel Pereira da Silva.

They assumed that all the stakeholders of the association convey the message that the law of euthanasia will be bad and are convinced that there are misconceptions that contribute to it being thought to be a good law when it is not.

They also stressed a second misconception which is the fact that the hypothetical right to kill oneself may arises from the fundamental right to self-determination and free development of personality. In addition, they mentioned the Universal Declaration of Human Rights, which considers the recognition of dignity to be inherent to all members of the human family and their equal and inalienable rights.

The third misconception is that we are faced with a law that will have safeguards but does not offer guarantees about how the criteria will be met and that monitoring of the respect for these criteria will exist since the bills do not specify it.

The fourth misconception stresses that euthanasia is not progress but regression.

The fifth is that euthanasia only affects the patients, while it also affects health professionals, the National Health Service, and families, among others.

They ended their speech by mentioning the sixth misconception of the bills, which is that the law is immune to the phenomenon of the 'slippery slope'.

6.3   Religious Groups and Religion-Oriented Professional Associations

6.3.1   *National Commission for Justice and Peace*

Besides the mentioned entities, the **debate of the first round of bills** also included a hearing with the National Commission for Justice and Peace, a 'secular' entity connected to the Portuguese Episcopal Conference, which was represented by its president, Pedro Vaz Patto. During the hearing, he stated that two fundamental principles are at stake: The inviolability of human life and the prohibition of killing since euthanasia is *"about causing* *death and not assisting in death".* He therefore ensured that the provision of end-of-life health care does not imply therapeutic obstinacy and that any life always has dignity and that this value is never lost.

Regarding individual autonomy and freedom, he stated that *"life is the presupposition of all rights (…) and autonomy cannot be invoked to justify the destruction of what is the root of one's autonomy".* He also considered that euthanasia *"cannot be a response to suffering, [since] with the death caused, suffering is not eliminated, the person who suffers is eliminated".* Finally, he mentioned the examples of the Netherlands and Belgium to comment on the increase cases of euthanasia, the so-called phenomenon of the 'slippery slope'.

6.3.2   Association of Portuguese Catholic Physicians

One of the interest groups which participated in **the second round debate** was the Association of Portuguese Catholic Physicians—which we consider not as an expert but as a religious interest group, since the Order of Physicians had been heard before—in a hearing held on July 1, 2020, in which the association was represented by two physicians—Luís Mascarenhas Lemos and Margarida Neto.

During their hearing, they reaffirmed their opposition against the decriminalization of euthanasia and presented their arguments for their position, stating that *"(…) we are physicians, we want to honour and comply with our code of honour, which we understand as a guarantee of respect for human life, from conception to natural death".* According to them, this is the position of all representatives of the association and even of the Order of Physicians, including the current and former presidents of the Order.

They stated that the principles of medicine exclude the practice of euthanasia, dysthanasia and assisted suicide and that *"medicine cannot be instrumentalized, with objectives that are unrelated to its activity, its practice, its ethics, and its fundamental law".* Therefore, they considered that physicians' role is to reduce the suffering of the patient, with their technique and humanity, and that *"(…) it is not possible to be a physician without going through the confrontation with suffering and death".*

In this regard, they considered that physical and intense pain, even anguish itself, could be relieved with medicines and other therapeutic forms, thus enabling the patient's well-being and absence of pain, even in the terminal phases of a given disease. They also stressed that 'intolerable suffering' cannot justify death or justify death on request, and that this would be *"the death of medicine itself, of the act of the caretaker"*. They also demonstrated their position against dysthanasia, just as they are against euthanasia.

For these reasons, they <u>advocated for extending palliative care and continuing care networks</u> and said that physicians need more training in this regard, which they considered insufficient. In addition, they claimed that public policies are needed to protect the weakest, such as the elderly and lonely people, justifying that people who do not feel alone do not ask for euthanasia.

At the end of their intervention, they revealed concern about the 'slippery slope', mentioning the Netherlands as an example. Finally, they stated that euthanasia would never be a medical act and that it was important to look at the opinions of technical entities and experts, especially the National Council of Ethics for Life Sciences.

6.3.3   Interreligious Working Group on Religion & Health

Like the previous entity, the **second round of debates** also counted with a hearing held on July 1, 2020, from the Interreligious Working Group on Religion & Health, which was represented by Fernando Sampaio, representative of the Catholic Church, and Paulo Pedro, representative of the Portuguese Evangelical Alliance.

During this hearing, they declared that their working group was supported by the position of 18 signatories, with different religious traditions, and that they signed a joint declaration in 2018. They also stated they would not speak about the bills, but rather about the legislative process and their concerns, which they expressed through the following sentence:

> *(...) every human being is bearer of an intrinsic dignity, which predates any criterion of autonomy and freedom or quality of life, and which does not depend on the stages of life through which they pass, nor on their conditions, on the social roles they play, nor on the culture to which they belong.*

In this regard, they stressed that the pandemic period associated with COVID-19 taught the lesson that society should be oriented to the defence of life. Moreover, they considered that there are no unworthy lives, despite going through life difficulties, diseases, and sufferings, and that it is proper for people to heal and care for others who are sick and fragile. In this sense, they thanked medicine for the following patients, especially regarding palliative care, even if they consider it to be insufficient, as well as the network of continued care and support for the elderly.

Under these circumstances, they believed that decriminalizing euthanasia would give the idea that *"(…) there are dignified and socially useful beautiful, autonomous lives, and others are unworthy and ugly, unpleasant, dependent and therefore useless, and therefore disposable"*. They assumed that euthanasia offers suicide and homicide as the only way out of suffering. They considered that, for their religious traditions, life is a gift, with a sacred character, and those religious qualities only confirm the intrinsic dignity of nature, which life contains, from the beginning to its natural end.

In this regard, at the end of their speech, they presented three evils that will arise with the approval of euthanasia: mistrust, injustice and a new duty. Regarding 'mistrust', they stressed the commandment of God's law, "you shall not kill, others or yourself", which is a source of trust in social coexistence. On 'injustice', they considered that the approval of euthanasia will bring the message that there are lives which are dignified and others are not. Finally, the 'new duty' emerges through the new social and family pressure on the disposable and more fragile, inducing the right/duty to ask for death.

6.3.4   *Portuguese Association of Catholic Jurists*

On July 8, 2020, the Portuguese Association of Catholic Jurists was represented in a hearing by Domingos Freire de Andrade Pedro Vaz Patto. In this hearing, they declared that, unlike what they usually do, they would question the meaning of the law and not the technical details, because they considered that it is not possible to apply the law only to exceptional cases, as shown by the phenomenon of the 'slippery slope' in other countries.

They considered that euthanasia is an issue of conscience and that with its legalization *"two fundamental foundations of civilization and the legal order are affected"*. These foundations are the inviolability of human life and the principle that human life is worthy of protection at all stages.

At the end of their hearing, they asserted that the bills will open the door to the 'slippery slope', as happened in other countries where active euthanasia is legal.

6.3.5   Portuguese Union of Seventh-Day Adventists

On July 18, 2022, that is, during the **fourth round of discussions**, the Portuguese Union of Seventh-day Adventists sent its written contribution to the committee in which it recognized and appreciated the invitation as a church based in Portugal since 2007 and present in Portugal since 1904. They began by recognizing that it was not lightly that they took a position on a public matter, but it was with a sense of responsibility and to contribute to its ecclesiastical reflection on the issue of life that was about the value of life and in opposition to euthanasia.

They added that they were signatories to the Declaration of the Interreligious Working Group on Religion & Health on May 16, 2018, in which several religious organizations spoke out against the decriminalization of euthanasia through a joint declaration. In view of this, they made the following statement:

> (…) human life is a wonderful gift given by God and worthy of being protected and sustained; and human being, because of their origin, are unique, irreplaceable and endowed with intrinsic dignity, regardless of their social, or ethnic identity, gender or situation, including situations of suffering and proximity to death; that scientific and modern medical advances have provided, through the use of technological means, pharmacological and therapeutic conditions, the minimization of suffering and the prolongation of quality of life; that faith, the spiritual dimension of the person, provides a mysterious force capable of helping to transcend, to find a meaning for suffering, a meaning for life and a hope that sustains existence in all its stages.

At the end of their written letter, they argued that legalizing freedom to ask for one's own death is admitting that they have failed and that the mission of the Seventh-Day Adventist Church is to convey hope and purpose to life.

6.3.6   Portuguese Buddhist Union

The **Portuguese Buddhist Union** also issued a written statement on July 18, 2022, in which it considered that the debate was insufficient and that the different spiritual and religious perspectives had not been included. They also considered that the new definitions of the bills have not been sufficiently developed, since, for example, the difference between the act of abstaining from prolonging life and the act of hastening death is not addressed. They added that the bills under discussion do not privilege life, physical relief, psychic, emotional and spiritual suffering and should <u>provide for the development of continued, palliative and terminal care</u>.

In addition, they stated that the conscientious objection stipulated in the bills covers only physicians and nurses, neglecting other health professionals and administrative staff. In the end, they stated *"(...) that the issue involves great complexity, with implications in the ethical, spiritual, social and behavioural aspects, which are not sufficiently studied and cautioned"*.

6.3.7   *Portuguese Association of Catholic Psychologists*

On September 15, 2022, the Portuguese Association of Catholic Psychologists submitted a written statement. In this statement, the association stated that the criteria for allowing euthanasia—serious and incurable disease, and definitive injury of extreme severity, which cause suffering and loss of dignity—are at stake.

About the 'suffering of great intensity', they stated that *"it results from an individual perception and feeling, influenced by numerous factors, highlighting mental health, emotional state, the quality of interpersonal relationships and satisfaction with living conditions"*. In this sense, they questioned whether physicians have the clinical competence to issue a reasoned opinion on mental health, emotional state, quality of relationships and satisfaction with the quality of living conditions. They also questioned whether physicians have the competence to assess decision-making capacity and central capacity determined by the patient's request.

On the concept of 'dignified death', they stated that *"the ill do not want to die, but want to stop suffering"*, and that dignity is not an attribute of death, but rather an attribute not alienable from life, the desire for death being against human nature. From this perspective, they reported that the bills entail the involvement of seven or eight people to access the patient's request, none of which are specialists in the evaluation and treatment of psychological and spiritual suffering.

At the end of their written pronunciation, they stressed the lack of alternatives that change and/or improve the quality of life of the ill and that, therefore, we are all failing as a society. They assumed that dignity is a value of life, and if there is no effective response to palliation and regular psychological intervention, bills are unsustainable.

6.3.8 The Lisbon Jewish Community

Finally, the **Lisbon Jewish Community** also issued its written statement, in which began by stating that nowadays many people live for many years in a state of terminal illness, unconscious, comatose, or in a vegetative state. They affirmed that the Jewish faith values life and its preservation above all the commandments of the Torah and that there are three biblical prohibitions, death being preferable to its transgression, idol worship, murder and forbidden sexual transgressions. For Judaism, human life is an absolute right and cannot be revoked by anyone, not even physicians. Under these circumstances, they mentioned that the practice of active euthanasia is prohibited in their religion and that *"merciful death is considered murder"*. About passive euthanasia, they recognized the need to relieve pain and suffering but prohibit all acts that shorten life.

NOTE

1. All the hearings, as well as requested and non-requested written opinions are available through the links for the bills and petitions provided in Chap. 3. The document whose link is provided in the following footnote also contains direct links.

CHAPTER 7

The Extra-Parliamentary Public Debate on Euthanasia: Positions Expressed by the Top Echelon of the Portuguese Catholic Church and in the Written Media

Abstract In this chapter, we analysed the extra-parliamentary debate on active euthanasia, namely the positions expressed by the top echelon of the Portuguese Catholic Church and by several actors in the written media.

Regarding the top echelon of the Portuguese Catholic Church, it was possible to observe that the Portuguese Episcopal Conference produced several notes expressing their opposition to the legalization of euthanasia, condemning the approval of different bills and congratulating the vetoes from the Constitutional Court and the President. Moreover, it is also important to highlight that, in 2018, when the first round of votes was about to take place, the Catholic church decided to sign a Joint Declaration of Religious Confessions of the Interreligious Working Group on Religion & Health against euthanasia and to organize the so-called 'Week of Life', during which information leaflets on euthanasia were distributed.

As concerns the written media, we identified 169 opinion articles published in one of the most reputed mainstream newspapers (Público) between January 2016 and May 2023, particularly in months which were crucial in the legislative process. These opinion articles were mainly against the legalization of euthanasia, while the most active authors were the scientific community, political actors and health professionals.

© The Author(s), under exclusive license to Springer Nature Switzerland AG 2023
I. S. Almeida, L. F. Mota, *Politics and Policies in the Debate on Euthanasia*, https://doi.org/10.1007/978-3-031-44588-0_7

Keywords Euthanasia; Portuguese Catholic church; Written media; Opinion articles and editorials; Positioning

To study the debate on euthanasia solely based on what happened in Parliament would be incomplete, hence the need to also analyse extra-parliamentary public debates. As it would be impossible to analyse all the public fora, we decided to focus our analysis on two: on the one hand, the debate in the Portuguese Catholic Church, namely by analysing public statements from the Portuguese Episcopal Conference; and, on the other hand, the media, namely by analysing the opinion articles in one main-stream newspaper (*Público*).

7.1   THE PARTICIPATION FROM THE TOP ECHELON OF THE PORTUGUESE CATHOLIC CHURCH

The public debate related to the issue of euthanasia was also marked by the participation of the Portuguese Catholic Church. To understand its participation in the debate, we analysed the documents, such as pastoral notes, published by the Portuguese Episcopal Conference, which is the group of bishops of the Roman Catholic Church in Portugal and is one of the most important structures of the Catholic Church in the country. The choice to analyse the Catholic Church derives from the fact it is the religion of most of the Portuguese population.

Following Vatican guidelines, the Portuguese Catholic Church condemns all kinds of euthanasia, and there were several moments when it revealed this position publicly.

One example was the **final communication of the Plenary Assembly** of the Portuguese Episcopal Conference,[1] held in April 2009, in which the representatives of the Portuguese Catholic Church stated that *"to die with dignity is to die with greatness and generosity, accepting suffering in its positive and redemptive dimension"*.

Later, in 2016, the Portuguese Episcopal Conference issued a **pastoral note** titled "Euthanasia: what is at stake? Contributions to a serious and humanizing dialogue",[2] in which they recalled one of the commandments of God's law, "thou shalt not kill", thus considering unacceptable the practice of euthanasia or assisted suicide. This entity also affirmed that, for believers, life is a gift of God, and the intrinsic value of human life in all its phases is rooted in our society and with a Christian brand, also recognized by the Constitution of the Portuguese Republic. Moreover, the pastoral note also stated that *"human life is the presupposition of all rights and all earthly goods, it is (…) the presupposition of autonomy and dignity"*, and that the request for euthanasia can never be guaranteed in what is a free, unequivocal and irreversible request, often reflected in a spontaneous moment.

In 2018, the Portuguese Catholic Church signed the **Joint Declaration of Religious Confessions of the Interreligious Working Group on Religion & Health**,[3] together with other Portuguese religious organizations such as the Evangelical Church, the Hindu community, the Islamic community, the Jewish community, the Buddhist community and the Ecumenical Patriarchate of Constantinople, in which they declared themselves to be against this practice. In this statement, it can be read that *"(…) every human being is unique and, as such, irreplaceable and necessary to the society of which they are a part, subject of an intrinsic dignity prior to any criterion of quality of life and utility, even to natural death"*. In addition, they believe that palliative care is the most complete achievement of the possible response by the State since it combines scientific and technical skills with compassion.

Together with the joint declaration, the Catholic Church strengthened its position with the **publication of an information leaflet on euthanasia**, distributed during the 'Week of Life' in May 2018, when the bills of the first round were being discussed and voted on. In this flyer, the Catholic Church presented euthanasia as a civilizational setback, also presenting the risk of the 'slippery slope' and the destruction of the doctor-patient relationship.

In 2019, the Portuguese Episcopal Conference presented a **new pastoral letter**,[4] in which they again discussed the issue of euthanasia, also stating that it was against the inviolability of human life and that euthanasia had been rejected by Parliament in the previous legislature. In this letter, the Church again supported its position with the experience of other European countries where euthanasia is legal, such as the 'slippery slope', and that continued and palliative care is not sufficiently promoted by the State.

In 2020, when the possibility of re-discussing euthanasia in Parliament became known, the Permanent Council of the Portuguese Episcopal Conference issued a **statement**[5] recalling the pastoral note issued in 2016, reinforcing that palliative care is a worthy option against euthanasia and stated, in the end, that it was monitoring and supporting *"the ongoing initiatives against the decriminalization of euthanasia, holding a referendum"*. At the time of the approval of the five bills in 2020, the Episcopal Conference issued a brief note[6] stating that it was "with great sadness" that they witnessed the approval of the bills, citing Pope Francis who "(…) called on health professionals to «constantly take into mind the dignity and life of the person, without any effect on acts such as euthanasia, assisted suicide or the suppression of life, even if the state of the disease is irreversible".

Also in 2020, after the proposal to organize a referendum on the decriminalization of euthanasia was rejected, they lamented Parliament's decision, saying it was *"the last way (…) to defend something that we deem essential and civilizational, it is not just a matter of the Church"*.

Later, after the decision of the Constitutional Court in 2021, the Episcopal Conference welcomed this decision, stating[7] that any decision in favour of the decriminalization of euthanasia and assisted suicide would always be contrary to the dignity of the human person and the Constitution of the Portuguese Republic.

More recently, the Episcopal Conference Portugal reacted against the new four bills, underlining its opposition to the legalization of euthanasia in a statement issued in June 2022,[8] recalling the commandments of the law of God, stating that *"when the law of men allows the State—sometimes and in certain cases—to take life, we are all exposed"*. On this note, the Episcopal Conference maintained that the new bills under discussion already present an extension of cases by covering situations of incurable disease and disability, joining the situations of imminent death discussed in previous projects. In addition to several arguments already put forward in previous notes, they added the danger that euthanasia and assisted suicide pose to the most vulnerable, *"who feel socially pressured to require euthanasia, because they feel they are "a burden" on family and social terms"*. At the end of the declaration, they mentioned that they hoped that this legislative process would not go further by political means, but by the choice of citizens.

In addition, on December 7, 2022, the Episcopal Conference of Portugal issued a note[9] regarding the approval of the bills, where they claimed to receive the news with sadness and hoped that the approved bill could still be changed. As in the previous notes, they recalled that the diploma did not guarantee the right balance between the protection of life and respect for the patient's autonomy, constituting a grave threat to humanity.

On December 13, 2022,[10] they made a statement to the Permanent Council, that they again expressed sadness at the approval of the legalization of euthanasia, showing harmony with other civil, Catholic and other religious institutions that manifested themselves in the same field.

7.2 Opinion Articles Published in One Mainstream Newspaper

The public debate on the issue of euthanasia was examined through a content analysis of opinion articles published on the website of one Portuguese generalist newspaper named *Público*, which is one of the most reputed daily newspapers and known for having a central political position. We opted to analyse only opinion articles, as they entail a clear intent from their authors to try to influence public opinion. This content analysis was based on three categories: the type of author (politicians or former politicians; health professionals; representatives from civil society associations/social movements; members of the scientific community; religion-related actors; jurists/lawyers; regular journalists and columnists; and ordinary citizens)[11] and the positions regarding euthanasia (positive, negative or neutral/uncertain).

Our analysis revealed that 169 opinion articles (which covered the topic of euthanasia in a significant part of the article) were published between January 2016 and May 2023. As can be seen in the following Fig. 7.1, the distribution of these articles throughout time was uneven and strongly related to political events described in Chap. 3, which reveals a clear intention by the authors of these texts of trying to express their opinions and eventually to influence decision-makers and/or public opinion. The first peak of articles was achieved in March 2016—the month before the first petition was submitted and while signatures were being collected—when nine articles were published. Some months later, a new peak of seven opinions articles was registered in February 2017, shortly after the second petition was submitted and the first bill was submitted. The third peak came with 23 opinion articles published in May 2018, when the first-round bills were discussed and voted on. The fourth and highest peak of 37 articles was achieved in February 2020, when the five bills of the second round were discussed and approved. The fifth peak (of 10 opinion

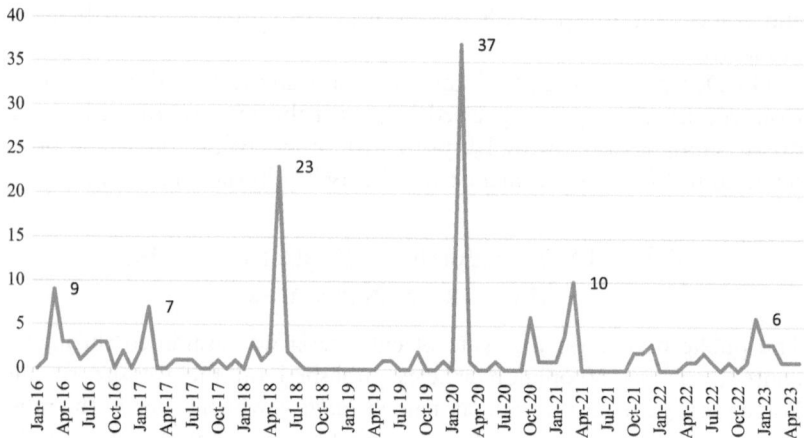

Fig. 7.1 Evolution on the number of opinion articles about euthanasia published on the website of the newspaper *Público*. (Source: own production)

articles) was registered in March 2021, when the Constitutional Court declared the second-round bill to be unconstitutional and the President vetoed it. Since then, no other peak has been registered, apart from the one in December 2022 (six opinion articles). A general view reveals that the discussion on euthanasia through opinion articles published in this newspaper was particularly active during the first and second rounds.

The content analysis (see Table 7.2) enabled us to verify that the categories of actors who published most opinion articles were the scientific community, particularly university lecturers ($n=40$; 23.67%), political actors ($n=35$; 20.71%), health professionals ($n=34$; 20.12%) and representatives from civil society associations/social movements ($n=23$; 13.61%). On the other hand, we concluded that actors in the media who intervened least were religion-related actors ($n=5$; 2.96%) and 'ordinary citizens' ($n=4$; 2.37%).

Moreover, it was possible to conclude that the majority (50.89%) of the opinion articles expressed a negative opinion regarding the decriminalization of euthanasia. These negative positions towards euthanasia were particularly expressed by health professionals and politicians, categories of actors in which the number of articles with positive positions towards euthanasia was much smaller. These figures may be explained by the fact that euthanasia is a topic that is deeply related to healthcare professionals

and their ethical codes, but also by the fact that most healthcare professional associations were against the legalization of euthanasia. Regarding
politicians, the high figures of negative opinions may be related to the fact
they wished to express their opinions outside Parliament, where the positive opinions were a majority. Moreover, the legislative process analysed in
Chap. 3 was a rather political one.

On the other hand, there is a balance of positions among the articles
published by civil society associations/social movements (12 in favour of
euthanasia vs. 11 against). The only category of actors in which the number of articles supporting euthanasia is significantly higher than the number of articles against it is that of the scientific community (19 in favour vs.
14 against) (Fig. 7.2).

As concerns the scientific community, the identified list of opinion articles was written by different authors, including lecturers and researchers in
the domains of law, sociology, philosophy, history, geography, bioethics,
etc. The author who wrote the greatest number of articles ($n = 8$) was
André Lamas Leite, a university lecturer in law and a regular writer in this
newspaper, who wrote an opinion article in February 2020, when the five

Category of actors	Nr. of opinion articles (and %)	Positive position	Neutral/ uncertain position	Negative position
Scientific community	40 (23.67%)	19	7	14
Political actors	35 (20.71%)	15	-	20
Health professionals	34 (20.12%)	6	4	24
Civil society associations/ social movements	23 (13.61%)	12	-	11
Regular journalists and columnists	14 (8.28%)	3	4	7
Lawyers and judges	14 (8.28%)	7	2	5
Religion-related actors	5 (2.96%)	1	2	2
Ordinary citizens	4 (2.37%)	-	1	3
Total	**169**	**63 (37.28%)**	**20 (11.83%)**	**86 (50.89%)**

Fig. 7.2 Opinion articles published in the newspaper *Público*, by type of actors
and advocated position. (Source: own production)

bills of the second round were discussed and approved, from which the following is excerpt:

> *A vote of yes does not require anyone to resort to euthanasia but reinforces a right. Yes, a sad right, of course, as none of us will be happy with the news that others felt that the only way out of their incurable and fatal illness was to hasten death. So many supporters of no, from the top of their morality, between the lines accuse the defenders of yes of being a kind of walking gas chambers. Disinformation, divine punishment, and foreign evil practices, with which no one can agree, are flags used by those who understand they have a right over the lives of others. The right to impose on someone who suffers atrociously because only in this way will the "social order" remain the same.*
>
> *André Lamas; university lecturer in law (scientific community); February 17, 2020*[12]

With a contrary opinion, we can highlight an opinion article published by the Institute of Bioethics of the Portuguese Catholic University in a very early stage of the debate when only the first petition had been submitted. The following excerpt from this opinion article is a good example of the arguments used against the decriminalization of euthanasia.

> *(…) euthanasia (death caused by a person at the request of the person who is killed) and assisted suicide raise far more serious problems that must be looked into. The defence of these interventions invokes the principle of autonomy to justify them, according to which we would be free to decide when to die. It cannot be ignored that this argument, by absolutizing the right to dispose of oneself, overvalues the concept of autonomy, which is always relative, limited by the freedom and interests of others (private and public), and subject to influence and coercion. (…) We must strive for an assisted, dignified death, without pain or suffering, accompanied and, whenever there is an indication, with palliative care.*
>
> *Institute of Bioethics of the Portuguese Catholic University; scientific community; November 1, 2016*[13]

Another category of actors which had a very significant participation in the public debate was that of political actors. In this group we can identify both MPs from Parliament—namely those who represented their political parties in presenting and publicly discussing the bills, such as José Manuel Pureza (BE), Pedro Filipe Soares (BE), André Silva (PAN), Isabel Moreira (PS) and Alexandre Quintanilha (PS)—but also other political actors who were not MPs at the time they published their articles—e.g. José Ribeiro e Castro (former leader of the CDS-PP), António Bagão Félix (former MP from the CDS-PP) or Paulo Rangel (former MP from the PSD).

In these opinion articles, it is possible to see some more ethical arguments being discussed, but also some political arguments being raised. This is the case of the two excerpts below, which use the arguments of lack of political legitimacy for political parties to submit the bills, or advocate for the organization of a referendum considering the lack of public debate.

> *Euthanasia entered the public debate in this legislature. In a great hurry. And there are good reasons to hope that it would not be so. There is no record of the theme having been covered in the 2015 campaign. There is no memory of a proposal that a party had previously presented.*
>
> *In other words, there is not the slightest democratic legitimacy to precipitate legislative decisions on such a fundamental, decisive and final matter. If anything can be said about the initiatives presented, which will be debated on May 29, it is that they do not result from any democratic mandate. On the contrary.*
>
> *José Ribeiro e Castro; lawyer and former leader of the CDS-PP (political actor); May 19, 2018*[14]

> *The hurried and expeditious way in which some parliamentary groups or some parliamentarians want to legislate on the issue is also very shocking. And the horror or repulsion that the organization of a referendum causes on this particular issue is not entirely understandable. Indeed, is it acceptable, is it admissible, that a matter of this nature can be decided on a simple and normal schedule as if it was about a mere regulation? The dignity of the issue would not justify a major national campaign, a participatory and informed discussion, preceded by debates in the media and social networks, but also by "citizens' assemblies" inspired by the Irish model (which preceded the referendums on abortion and same-sex marriage)?*
>
> *Paulo Rangel; member of the European Parliament (PSD); February 11, 2020*[15]

The group of political actors who wrote opinion articles in favour of the decriminalization of euthanasia also raised several substantive arguments, but also more political arguments, such as responding to the criticisms about the quality of the debate or criticizing the President for vetoing the bills based on personal political/religious reasons rather than on juridical reasons.

More importantly, this debate was not raised from the top down. Paulo Rangel omits the Right to Die with Dignity—Civic Movement for the Decriminalization of Assisted Death" (did you forget?). This civic movement, which defended the decriminalization of assisted death in Portugal, was created five years ago, publishing a manifesto signed by several dozen personalities from the most diverse areas of society (allow me to recall João Semedo), which resulted in a petition, which was delivered to the Assembly of the Republic in April 2016 with around 8400 signatures and, contrary to what Paulo Rangel says, the debate never stopped.

The debate was intense and in no way lacking in society, in universities, in the media, in schools and, of course, in Parliament. The discussion in a working group that for months listened to jurists, ethics people, and physicians, among many other participants, were joined by bills from those who democratically understood to say yes to the petition. (...)

As I have already written in another text, on March 11, 2016, a poll was published in Expresso according to which 67.4% of Portuguese people are in favour of euthanasia. Despite this, we remained firm in refusing to have a referendum on a complex issue of fundamental rights.

Isabel Moreira; MP from the PS; February 13, 2020[16]

Twice, the Assembly of the Republic approved, by far more than an absolute majority of members of parliament—138 out of 230—the decriminalization of assisted death. (...)

In both cases, the President of the Republic resorted to his constitutional prerogative not to enact what Parliament had approved. First, he invoked doubts about constitutionality. He now invokes doubts about legal interpretation. (...)

Therefore, the late perplexities of the President of the Republic cause perplexity. It is perplexing that a political veto comes dressed as a veto for legal reasons. The timing and manner of justifying the presidential veto has an obvious political impact. And intended by its author. A more political veto than this would be difficult.

José Manuel Pureza; former MP of the BE; December 1, 2021[17]

Another category of actors which had a very significant participation in the public debate was that of health professionals, particularly active in advocating against euthanasia. Although they used all different kinds of arguments, the authors from this category of actors quite often used arguments related to the way they perceived the ethics of medical practice, the need to improve the healthcare system before decriminalizing euthanasia or the dangers of patients requesting euthanasia because of internal or external pressures. The following two excerpts—one from the president of the Order of Physicians and another from 'regular' physicians—are good examples of this.

> *As the president of the Order of Physicians, I am on the side of life. I defend it every day, in the exercise of my duties, in the application of the code of ethics, but also, as a physician, in the routine of health care, I provide to patients. This is how medicine is regulated, a practice that defends life—and the quality of life—of all people. (...)*
>
> *And we all have the right to health care with quality that allows us to evaluate, as patients, the options for having more life with the highest possible quality. (...) We all want to live. And live well. The health system needs to improve a lot before allowing itself to consider the decision of death. Life must be preserved, in the best conditions, with the greatest dignity. All the way to the end.*
>
> *Miguel Guimarães; president of the Order of Physicians (health professionals); May 29, 2018[18]*

> *In Portugal, a vote on bills to legalize euthanasia and assisted suicide is scheduled to take place shortly. What dangers does this patient face with the approval of these laws? (...)*
>
> *1. The patient feels and knows that his/her life becomes less valuable than others, with less dignity and, therefore, can be discarded.*
>
> *2. The family and the associations that support the patient receive a message of devaluation of the chronic and disabling disease. On the one hand, the patient can more easily be regarded as a nuisance and an obstacle; and, on the other hand, the lack of solidarity, family dysfunction and eventually opportunism and greed may push the patient towards the so-called "dignified death". (...)*
>
> *3. In a country where only 25% of the population has access to palliative care, it is to be expected that, when faced with a situation of worsening or progression of the disease, with greater fragility and suffering, the option of euthanasia.*
>
> *Francisco Adão da Fonseca; physician and expert in internal and intensive medicine; February 18, 2020[19]*

Although the list of articles written by health professionals who are in favour of decriminalizing euthanasia is short, we believe it is worth mentioning it, particularly the contribution from Rosalvo Almeida, a retired neurologist, who wrote five opinion articles (out of six) that support euthanasia. In these articles, he advocated, for instance, that palliative care is not a universal response to unbearable suffering or that the Hippocratic Oath is not entirely in contradiction with euthanasia.

The debate on medically hastened death has put its opponents in the position of advocates of palliative care, and it seems that those who defend this hastening are opposed to palliative care. This is a generalized confusion that nobody benefits from (…)

We know, from empirical knowledge and also from studies carried out, that many people in the terminal phase of their illness, even if they receive the best care, continue to ask to die—their suffering can be refractory physical pain or be due to existential reasons that no one has the right to contradict. (…)

But is it true or not that the Hippocratic Oath requires doctors to do everything for the lives of their patients?

Yes, health professionals must do everything to save them, but it also obliges them not to harm them. If the best palliative care cannot prevent the continuation of suffering, persisting with measures that, by prolonging life, only prolong suffering, that is harming.

Rosalvo Almeida; retired neurologist; September 5, 2016[20]

As expected, representatives from civil society organizations and movements have also written several opinion articles, either advocating the decriminalization of euthanasia or speaking against it. On the one hand, all the articles in favour were naturally written by representatives from the Civic Movement for the Decriminalization of Assisted Death (which submitted the first petition), namely Bruno Maia, João Ribeiro Santos and Laura Ferreira dos Santos, particularly the latter, who wrote nine texts in 2016, before her death from a prolonged disease in December 2016. The following excerpt is from an article in which she presented her arguments and criticized the opposing petition.

1. On May 14, a petition appeared in the Expresso [a highly reputed weekly newspaper] entitled "All life has dignity". Above all, it is a petition that asks the Assembly of the Republic to prohibit euthanasia (there is no mention of medically assisted suicide), thus implicitly opposing the previous petition of the Right to Die with Dignity Movement. In the meantime, the protection of the elderly and disabled, the most vulnerable in society, is also requested, thus insinuating that they could be murdered with the decriminalization of euthanasia.

2. Grounds for this petition? To consider, without any complexity, that "human life" is "inviolable, inalienable, inviolable". In fact, if there is this unavailability, if there is an obligation to live up to the end, I think that its promoters should also ask for the criminalization of suicide or anyone who attempts it, as has happened for centuries. Given that the promoters and signatories of this petition are fundamentally people who claim to be Catholic, I think it is not abusive to say that they want the State, although secular, to con-

tinue to enshrine in confessional law principles, even though there are well-known and respected Christians who are in favour of assisted death, such as the theologian Hans Küng, the former archbishop of Canterbury, George Carey, and the South African emeritus archbishop Desmond Tutu, among many others.

Laura Ferreira dos Santos; Civic Movement for the Decriminalization of Assisted Death and retired university lecturer; June 9, 2016[21]

More recently, Bruno Maia wrote two opinion articles, one of them dismantling several arguments against the decriminalization of euthanasia and presenting statistics that demonstrated that most physicians are in favour of it, concluding with the following excerpt:

But what cannot be left unsaid is that the majority of Portuguese people want these bills approved. Because they understand that this is the only way to respect the different worldviews that exist in a democracy. Because they know that it is their right to determine what dignity and autonomy are. And because they no longer live in darkness, they are not frightened by the apocalyptic announcements of change and are fed up with the paternalism of those who want to dictate them how to live their life… and their death!

Bruno Maia; neurologist and activist for assisted death; February 20, 2020[22]

On the other hand, representatives from the Portuguese Federation for Life—the proponent of the above mentioned petition against the decriminalization of euthanasia—have also used opinion articles to present their ideas and to respond to criticisms. For example, José Maria Seabra Duque wrote seven opinion articles throughout time, from one of which the following is an excerpt, a direct answer to the abovementioned article published by Laura Ferreira dos Santos.

1. On June 9, an article by Professor Laura Ferreira dos Santos, the main promoter of the "Right to Die With Dignity" movement was published in this newspaper. The article has the sole purpose of attacking the petition "All Lives Have Dignity" which is currently collecting signatures to be submitted to the Assembly of the Republic. (…)

In this article, the professor reduces the entire petition "All Lives Have Dignity" to a group of Catholics who want to impose their vision on society. For the pro-euthanasia activist, only Catholics are of the opinion that human life is inviolable, inalienable, and inviolable. Furthermore, she argues that defending these principles is against individual autonomy and is against democracy.

3. This preconceived view is born out of some obsession with the Catholic Church and ignores most of the Portuguese legal doctrine. If the professor had taken the trouble to study this issue a little more deeply, she would discover what any first-year law student knows: that personality rights, including the Right to Life, are, according to most of the doctrine, inalienable and inviolable. (...)

4. Of course, this view is debatable. In fact, just as any 1st-year law student learns that the right to life is an inviolable right, he/she also learns that on any subject of law, doctrine diverges. And it is an extremely interesting debate. (...)

5. What we are discussing now is whether the State, when faced with someone who asks to die, can decide whether to do so or not. And this is the big question: can the State judge the moment when life ceases to be dignified? Furthermore, can the State, through any health professional, kill a person? Can the censorship of homicide or aiding suicide be removed by the power of the State?"

José Maria Seabra Duque, subscriber of the petition "All Lives Have Dignity" and later signing as a board member of the Portuguese Federation for Life; July 14, 2016[23]

The entire euthanasia legislative process is difficult to understand. It is difficult to understand the rush to legislate a law that was rejected two years ago in Parliament, the insistence on legislating against the opinion of the Order of Physicians, the Order of Nurses, the Order of Lawyers, the National Commission of Ethics for Life Sciences and of so many other institutions and specialists, the obstinacy on a subject that has given rise to a huge wave of social contestation and without any relevant social movement that is favourable to it (...).

José Maria Seabra Duque, board member of the Portuguese Federation for Life; January 11, 2021[24]

Unlike what happened with other groups of actors, there were only a few opinion articles published by religion-related articles ($n = 5$). Unsurprisingly, two articles against the decriminalization of euthanasia were written and two others had an uncertain position, although one of them expressed the need to hear the opinion of experts such as Miguel Oliveira da Silva, former president of the CNECV, who is against euthanasia. The following excerpts reveal the positions of these actors, who mentioned, for instance, the danger of the 'slippery slope'.

Theo Boer [lecturer in bioethics and former member of committees to monitor the implementation of the Dutch law which allows euthanasia] said that in the Netherlands, in the early nineties of the last century, when euthanasia began to be legally practised in specific cases, no one thought that this would result in the situation that is now visible to all: an organized system of provoked death (in some regions, it accounts for around 14% of the total number of deaths). The breach that was opened has contributed to eroding the cultural edifice that rested on the foundation of the prohibition against killing. And it will continue to do so incessantly.

It is in all this, in the deepest, longest-term, and most far-reaching consequences, that whoever has the responsibility to legislate on euthanasia should consider.

Pedro Vaz Patto; president of the National Commission Justice and Peace and Judge; May 28, 2018[25]

The only opinion article published by a religion-related actor which expresses support for the decriminalization of euthanasia was that published by Joel Lourenço Pinto, President of the Synod of the Evangelical Presbyterian Church of Portugal, in which he explained the reasons for not signing the inter-religious declaration against the legalization of euthanasia (see Sect. 7.1).

The current debate on euthanasia and assisted suicide has recently allowed the different religious traditions present in our country to offer us a manifestation of interreligious unity by speaking out on the sacredness of human life (…).

However, by limiting the practice of mercy and compassion to palliative care (…) directly opposing assisted death in any of its forms, the statement is not without a certain perplexity. It was for this reason that the Presbyterian Church of Portugal whose Synod I have the honour to preside over, did not subscribe to this declaration.

As Christians and citizens, we believe that human life is inviolable and that the mission of the State is to do everything to protect it. We also share the call for the generalization of palliative care as a response to the end-of-life problem. However, despite being aware of their Christian identity, Protestant ethics understands that it has, nevertheless, an ethics of dialogue or in dialogue, allowing Christians to affirm their convictions, listen, and try to understand the convictions of others in a constructive way. For these reasons, we believe it is necessary not to forget those who consider that assisted death is not a mere murder but the last affirmation of their human dignity … a God who forbids a man from ending his life when he forces him to carry unusually unbearable burdens, such a God, is not his friend.

Joel Lourenço Pinto; President of the Synod of the Evangelical Presbyterian Church of Portugal; May 29, 2018[26]

Summing up, concerning the arguments in favour, the actors present, above all, respect for the principle of autonomy and human dignity and the right to death in patients whose pain and suffering are extreme and unbearable. In the field of arguments regarding medical care, they state that the mission of the National Health Service is to support users and that the support of palliative and continued care does not clash with the decriminalization of euthanasia. Moreover, they claim that the Portuguese State is secular and that its fundamental role is to defend rights, freedoms, and guarantees.

On the other hand, as concerns the argument against euthanasia, several actors present the need to strengthen the National Health Service and palliative and continued care. Moreover, they mention the defence of the Hippocratic Oath and the fact that the physician-patient relationship would be impaired if euthanasia is legalized. They present the danger of the 'slippery slope', mentioning the examples of countries where active euthanasia is legal. They add that the role of the State is to protect the lives of citizens, stating as their main argument, the right to life stipulated in the Portuguese Constitution.

NOTES

1. See: https://www.conferenciaepiscopal.pt/v1/comunicado-final-da-assembleia-plenaria-da-conferencia-episcopal-portuguesa-6/
2. See: https://www.conferenciaepiscopal.pt/v1/eutanasia-o-que-esta-em-causa-contributos-para-um-dialogo-sereno-e-humanizador/?highlight=eutan%C3%A1sia
3. See: https://www.conferenciaepiscopal.pt/v1/declaracao-comum-das-confissoes-religiosas-sobre-a-eutanasia/?highlight = eutan%C3%A1sia
4. See: https://www.conferenciaepiscopal.pt/v1/um-olhar-sobre-portugal-e-a-europa-a-luz-da-doutrina-social-da-igreja/?highlight = eutan%C3%A1sia
5. See: https://www.conferenciaepiscopal.pt/v1/eutanasia-comunicado-do-conselho-permanente-da-cep/
6. See: https://www.conferenciaepiscopal.pt/v1/despenalizacao-da-eutanasia-breve-nota-da-conferencia-episcopal-portuguesa/
7. See: https://www.conferenciaepiscopal.pt/v1/comunicado-do-conselho-permanente-da-cep-face-a-aprovacao-da-eutanasia/?highlight = eutan%C3% A1sia
8. See: https://www.conferenciaepiscopal.pt/v1/legalizacao-da-eutanasia-e-do-suicidio-assistido/

9. See: https://www.conferenciaepiscopal.pt/v1/nota-da-conferencia-episcopal-portuguesa-sobre-a-aprovacao-da-legalizacao-da-eutanasia-e-do-suicidio-assistido/

10. See: https://www.conferenciaepiscopal.pt/v1/comunicado-do-conselho-permanente-da-cep-4/

11. When authors could be classified in two categories, we opted to choose the category which seemed more relevant for their position about euthanasia—e.g., José Maria Seabra Duque signs his articles as jurist and member of the Portuguese Federation for Life, and Bruno Maia signs his articles as physician and member of the Civic Movement for the Decriminalisation of Assisted Death. In both cases, we considered that from both of their positions were more influenced by their membership of civic movements or civil society associations, rather than by their professions per se.

12. See https://www.publico.pt/2020/02/17/opiniao/opiniao/radicalidade-humano-eutanasia-1904391

13. See https://www.publico.pt/2016/11/01/sociedade/opiniao/eutanasia-e-suicidio-assistido-ajuda-ou-ameaca-1749512

14. https://www.publico.pt/2018/05/19/politica/opiniao/licenca-para-matar-1830015

15. See https://www.publico.pt/2020/02/11/politica/opiniao/eutanasia-defesa-referendo-1903622

16. See https://www.publico.pt/2020/02/13/politica/opiniao/eutanasia-paulo-rangel-referendo-1903816

17. See https://www.publico.pt/2021/12/01/opiniao/opiniao/tao-politico-veto-1987102

18. See https://www.publico.pt/2018/05/29/politica/opiniao/tempo-de-viver-1832399

19. See https://www.publico.pt/2020/02/18/impar/opiniao/eutanasia-vidas-indignas-1904397

20. See https://www.publico.pt/2016/09/05/sociedade/opiniao/o-espantalho-e-outras-falacias-1743175

21. See https://www.publico.pt/2016/06/09/sociedade/opiniao/proteger-a-santidade-da-vida-mas-nao-a-dignidade-de-quem-morre-1734532

22. See https://www.publico.pt/2020/02/20/opiniao/opiniao/eutanasia-1904948

23. See https://www.publico.pt/2016/07/14/sociedade/opiniao/proteger-a-inviolabilidade-da-vida-humana-um-dever-de-cidadania-1738184

24. See https://www.publico.pt/2021/01/11/politica/noticia/eutanasia-birra-vai-longe-1945894

25. See https://www.publico.pt/2018/05/28/sociedade/opiniao/a-brecha-aberta-pela-eutanasia-1832054
26. See https://www.publico.pt/2018/05/29/sociedade/opiniao/por-uma-etica-da-transgressao-1832084

CHAPTER 8

Advocacy Coalitions in the Debate on Euthanasia

Abstract In this chapter, we analysed the perspective of interviewed political and social actors on the dynamics of agenda-setting of this issue, as well as on the dynamics of coordination among actors during the formulation process.

We concluded that, while the agenda of this issue had been set formally by the civic movement 'Right to Die with Dignity', the left-wing parties which submitted bills in the first round were already planning on doing it. Moreover, the topic remained on the political agenda even after a rejection in 2018, mostly due to those left-wing parties.

As concerns the relations among political parties, the interviewees considered that there was an easy coordination among political parties which submitted bills, something that was particularly visible in the process of combining different bills into conciliatory texts. Contrarily, political parties against the decriminalization of euthanasia did not create a particular coordinated strategy other than some occasional conversations.

Regarding the relations between political parties and civic movements, the interviewees considered political parties and civic movements in favour of legalizing euthanasia developed a close collaboration, even if some consider that the former overshadowed the latter. Contrarily, only sporadic conversations between political parties and civic movements against euthanasia were reported.

I. S. Almeida, L. F. Mota, *Politics and Policies in the Debate on Euthanasia*, https://doi.org/10.1007/978-3-031-44588-0_8

Keywords Euthanasia; Advocacy coalitions; Political parties; Interest groups; Interviews

After analysing the positioning and main arguments presented by the actors, in this chapter, we analyse the perspective of political and social actors on the dynamics of agenda-setting of this issue, as well as on the dynamics of interaction and strategic coordenation between actors during the formulation process. To this end, we interviewed representatives of five parties (two in favour of the legalization and three against) and two social movements (one in favour and one against) in May and June 2022.

8.1   Coalition Strategies
in the Agenda-Setting Process

One of the dimensions of our analysis concerns the agenda-setting of the legalization of active euthanasia in Portugal. In this context, we questioned the political actors about which factors, in their opinion, most contributed to its this.

For the representatives of the political parties in favour of the decriminalization of euthanasia which submitted bills, the petition presented by the civic movement 'Right to Die with Dignity' was of great importance in terms of agenda-setting. However, they assume that their parties would submit bills even if this petition had not been presented to Parliament. From this perspective, several parties assume that the issue of euthanasia has been on the political agenda since the discussion of the living will.

In this regard, the representative of the Left Bloc mentioned that, although they saw the civic movement as a trigger, their bill was in line with bills they had submitted in the past about other issues related to patients' autonomy in healthcare, as expressed in the following excerpt:

(...) the existence of the social movement capable of affirming a subject in Portuguese society is fundamental information for us (...). Therefore, the fact that a movement emerged in Portugal in favour of the decriminalization of assisted death with a wide scope, a great plurality, and force, with the capacity to dialogue with the most diverse sectors of society was decisive for the Bloc and other political forces to assume the responsibility of having bills on this issue. (...)

The reasons that led us to take this position and to be a party that proposes a bill on the decriminalization of assisted death, stem from a path that the Bloc set for its parliamentary intervention, firstly regarding end-of-life issues and regarding issues of patients' autonomy within the scope of the right to health. In other words, the Bloc was at the origin of a series of measures that were adopted over time, tending to increase respect for patients' autonomy. The case of the living will is one example. Combating therapeutic obstinacy is another such example. (...) Therefore, over time, the Bloc has intervened in this matter with initiative, placing itself in a prominent position in these initiatives.

This is followed precisely by the step of affirming this principle of respect for autonomy also regarding early death. Therefore, it is an additional step on the path of several others that had been taken before.

Interviewed representative of the Left Bloc

With a similar position, the interviewed PEV representative also stated that, regardless of the petition presented in favour of decriminalization, PEV would present its bill and, in this sense, declared the following:

We first discussed this in the party's bodies, to take the position, of course. It was in the National Council, (...) and what was decided next in the direction of the party was to create a working group to develop a bill to decriminalize medically assisted death, because that is what we are talking about, an amendment to the Penal Code. (...) When we drafted the bill, we did not hold hearings, but we read and studied all the results of the hearings that were held in the Assembly of the Republic, of which there were many. So, we didn't need to listen, we knew the positions of virtually every organization, whether for or against. (...)

Even in 2015, we considered that there were conditions to go forward. It was in this sense that we moved forward with the projects. (...)

But I think that even though we knew, and now I am speaking from the Greens' point of view, that the project would be rejected in the XIII Legislature [2015–2019], that would not prevent us from submitting the bill anyway and even putting it on the [Parliament's] agenda, even for discussion. We often submit projects that we already know, at the outset, are unlikely to go through, but there is also a discussion.

Interviewed representative of PEV

The parties that demonstrated their opposition to the decriminalization of euthanasia had different positions regarding the factors that contributed to its agenda-setting, particularly regarding the role played by civic movements. However, all of them stated that there was pressure from leftist parties to regulate euthanasia.

In this regard, the PCP representative highlighted the emergence of a 'euthanasia ideology', which brought together left and right political parties, and which had a liberal ideological matrix of recognizing this individual freedom, but which was not focused on an actual social problem. Regarding the civic movement, the interviewee considered that it was practically indistinguishable from political parties, as revealed in the following excerpt:

> *I think it was mainly an ideological question, from the point of view of a certain, let me use an expression, "euthanasia ideology". That is, can we consider that the non-existence of legislation that allows euthanasia in our country is a social problem that Portuguese society is confronted with? It does not seem so to us. In other words, even if this law is approved as proposed, will it have a significant impact, and will many people want to end their lives? We want to believe not.*
>
> *But what actually happened was a movement that was expressed in Portuguese society, which considered that legislating on this issue (...) would be a recognition of people's individual freedom to end their life. (...) When I talk about the ideology of euthanasia, I mean it in that sense, even in this liberal sense. It is no coincidence that these initiatives led to the convergence of sectors that come from the left and others from the right. The neoliberal right was also one of the authors of this initiative. (...)*
>
> *We have never considered this issue to be either a religious issue that divides people who are religious and those who are not, or an issued dividing left and right. (...)".*
>
> *I believe that we cannot speak exactly about a social movement, as this would imply a breadth and a movement that did not actually exist. There was a movement of opinion that arose in some political and media sectors, with strong media support, which helped a lot to give projection to this movement of opinion and to lead to bills being submitted. (...) The truth is that the authors of the bills came out of the promoters of petitions (...) I would say that these were processes that fed each other. I mean, there was in fact a movement of opinion which launched a petition which set the ground in the Assembly of the Republic that would then lead to the presentation of bills in line with that petition. But in fact, I believe it was the process of feeding on each other.*
>
> *Member of the PCP interviewed.*

For the representative of the Enough! party (CH), one of the main factors that contributed to the issue of active euthanasia being placed on the public and political agenda was the pressure from leftist parties, rather than civil society, as revealed in the following excerpt:

Clearly there has been permanent pressure from leftist parties, particularly from the extreme left, apart from the PCP, but with particular emphasis on the Left Bloc and PAN as well, and the left wing of the PS, namely on the part of MP Isabel Moreira. It is not civil society that is pressing. (…) If the extreme left is against popular consultation [through a referendum], it is because they are not very confident that popular consultation would give the go-ahead to euthanasia. There will be a group of people who will not be indifferent, but I think that in a popular consultation, with clear questions, euthanasia would be approved. (…)

In Portugal, civil society organizations have no weight. (…) They bring together half a dozen friends, on both sides, with an always with a high level of militancy. (…) But civil society in Portugal is not mobilizable. At least this is the conclusion that I have reached after several years of militancy both in party life and in civil society.

Member of CH interviewed.

The representative of the CDS-PP, in turn, stresses that the agenda-setting of this issue was based on the fact that the fight for the decriminalization of euthanasia had become an ideological flag, to which could be added the fact that the leaders of the civic movement had died, something that reinforced the emotional component of the fight.

From a political point of view, it is through an agenda that waves flags and that, far from solving the country's economic and social problems, goes to flags to generate an illusion. This process has been a process of illusions, generating an illusion that people are being offered things, a civilizational advance. (…)

If you consider the 2016 [civic] manifesto, there are very active people in this process who have already passed away, and some assumed that it was a tribute paid to them to be able to ensure that the process did not fall by the wayside. (…) And I personally believe that there are personal factors here that are linked to these circumstances, which led to the emotional component being reinforced, through the disappearance of several names,

There are flags here. There is a postponement of reality. It does not matter that people are not entitled to health care. What matters is that they may supposedly have another right.

Interviewed representative of CDS-PP

8.2   COORDINATION AMONG POLITICAL PARTIES

Through the interviews, it was also possible to analyse the extent to which strategies of interaction and coordenation between political forces would have been developed, either in favour or against the decriminalization of euthanasia.

8.2.1   Coordination Between Political Parties in Favour of Decriminalizing Euthanasia

The coordination between political parties was a particularly evident scenario among the political parties that submitted bills, which were mostly left-wing parties. This coordination took place, particularly through the process of creating conciliatory texts.

In the case of BE, the interviewee assumed that the process of coordination between bills was relatively simple, mentioning the following in this regard:

> It was fairly easy! Naturally, there were some differences and the method we followed was to reconcile the five projects in a single text. (...) It was quite easy to have a single text, maintaining, however, the possibility of autonomously voting on those solutions that were more specific to each political force, those differentiating solutions. For example, the Greens' bill provided that hastened death could only be carried out in National Health Service entities and not in the private or social sector. And, therefore, we understood that this proposal should deserve an autonomous vote, which ended up not being approved.
>
> Interviewed representative of BE

The interviewed PEV member had a similar perception regarding the coordination process between the various political actors in favour of decriminalizing euthanasia, as is evident in the following excerpt:

> Because we had a working group with elements from several political forces and everything went as expected. Naturally, in the working group one creates a text that tries to combine things, and here MP Isabel Moreira did a fantastic job, trying to create a text that encompasses the distinct concerns of the different bills. (...) But the work went very well, without major friction, without major problems.
>
> Interviewed representative of PEV

In this regard, we also questioned the political parties against euthanasia if they witnessed a process of coordination between the political parties in favour of euthanasia.

For the PCP representative, the positioning of the parties regarding this issue was not surprising, and he therefore considered that it was clear, from the beginning of the process, what the position of each political force was and that the voting result was not surprising. Concerning the coordination between political actors, the interviewee only mentioned having observed the coordination from a distance:

> *We, like everyone else, observed without participating directly in this process, because we understood that, not having a bill, it was the proponents who had the burden of coordinating with each other. So we waited for them to do so, and we expressed our opinion at the end.*
> *Interviewed representative of PCP*

In the same vein, the representative of CH mentioned that this party was also not surprised by the votes of each party, but just a little surprised by the position of the PCP, not so much because of their vote, but because of the eloquence with which they defended their arguments, as can be seen in the following excerpt:

> *For us, it was expected. (...) The position of the PCP was, for us, a surprise. I mean, a half-surprise, because the Portuguese Communist Party is, in many aspects, very conservative, apart from the economic part. In some social issues, it is quite conservative. But such a firm and well-founded position, as the one presented, came as a surprise to us. In the other parties, the CDS was expected, in the PSD the division was expected, the Left Bloc was expected, PAN, and so on.*
> *Interviewed representative of CH*

Finally, the interviewed representative of the CDS-PP mentioned that the rejection of bills in 2018 would have brought its proponents closer together since they were not expecting that outcome, which made them create a united front.

> *What I am saying here is that I believe that the political actors who proposed it did not think that the law would not be approved in 2018. They were not expecting that, and therefore had to 'close ranks'.*
> *Interviewed representative of CDS-PP*

8.2.2   Coordination Between Political Parties Against Decriminalizing Euthanasia

Regarding the parties that voted against the decriminalization of euthanasia, the interviewees reported that there had been some conversations, especially between PSD and CDS, but nothing very evident.

This idea was conveyed precisely by the interviewee from the CDS-PP, who admitted the existence of an instance of coordination between this party and the PSD in the scope of the first-round vote, in 2018, as is evident in the following excerpt:

> *In 2018, clearly. Obviously, a coordination, in this case, of the biggest parties, at the time, that were against it: the PSD and the CDS. (...) And therefore, yes, there was coordination, and there was coordination with civil society organizations against euthanasia.*
> *Interviewed representative of CDS-PP*

The representative of the Enough! party admitted that they had only some occasional conversations with other parties, particularly about the possibility of a request for successive review of constitutionality if the bills had not been rejected by the Constitutional Court and by the President, as is evident in the following excerpt:

> *There was nothing more than sporadic conversations, as far as I know. (...) But then there was the hypothesis that arose when the bill went to Belém [Palace¹]. (...) If, by any chance, the bill would be approved, without preventive review by the Constitutional Court requested by the President of the Republic, there was the possibility of gathering enough MPs who were against the approval of the law to later request a successive review. (...) I think there were one or two consultations from the PSD with André [Ventura]² about this, but there was no attempt to create a 'wall'.*
> *Interviewed representative of CH*

The PCP representative, on the other hand, said that this party only demonstrated its fundamental divergence regarding the legalization of euthanasia, and did not deploy itself together with other political parties to try to boycott the legislative process, as it is clear from this excerpt:

> *It was a legislative process, in which, after the bills were approved in general terms, their proponents drafted a common text with which they could relate,*

and PCP did not want to play an active role in this part. That is, not having been a proponent, it did not interfere directly or indirectly. (...) Therefore, we left that to those who had bills, and, after that, we then expressed our vote regarding the final text, which coincided with what we had had regarding the different bills, because, for us, the question was not in better or worse concrete drafts, but above all an underlying fundamental question.
Interviewed representative of PCP

In this regard, we also questioned the political parties in favour of euthanasia if they considered that there was some kind of coordination between the political parties with the opposite position. For the BE representative, the position of the political parties without legislative initiatives in this regard was not surprising, as it had already become clear during the extra-parliamentary discussion.

The PEV representative agreed with this assumption but also said he had not seen a 'destructive' approach by their opponents in the parliamentary committee.

I didn't see any manoeuvre there. I remember the participation of an MP from PSD, who also had a very urban attitude in the discussion.
Interviewed representative of PEV

8.3   COORDINATION BETWEEN POLITICAL PARTIES AND CIVIC MOVEMENTS

Besides analysing the coordination strategies among political parties, we also tried to understand if there had been coordinated strategies between parties and civil society movements, either in favour or against euthanasia.

8.3.1   Political Parties and Civil Society Movements in Favour of Euthanasia

In the interviews, the representative of the pro-euthanasia movement assumed that at an early stage, the heads of the movements were not associated with any political party. However, he assumed that after the presentation of the petition to Parliament, the proximity between the actors became evident. In this sense, he declared the importance of several party members in favour of the decriminalization of euthanasia, such as José Manuel Pureza from the BE, Isabel Moreira and Maria Antónia Almeida

Santos from the PS, and André Silva from PAN, but assumed that the relationship with IL was almost non-existent. Both ideas are expressed in the following excerpt:

> *The movement has three essential heads, unfortunately all three of whom have disappeared. The first was Laura Ferreira dos Santos, who already had a job in this domain in Braga and had no political party. The second was João Ribeiro dos Santos, who was a nephrologist in Lisbon and who had no party either. But these two people, who already had worked [in this domain] for many years, were separated, so to speak. The third head is João Semedo [former BE leader], who ended up putting these two people together, joining several other people, and creating a movement. So he was the organizer, he was the man who had the idea, he was the man who organized, he was the man who built this movement. And, right from the start, there were not even people from the Left Bloc, for example, in the movement. There was only João Semedo. The rest of most of the signatories, with an extensive network of contacts outside the Bloc, were person- alities mainly from the healthcare sector, but not only, who he managed to join to have a manifesto.*
>
> *I don't know how all legislative processes work, but I think that this process was different from others that I know of. It was different, because, right from the start, there was a great confluence of ideas and proposals. (...) There were no reservations from the parties regarding the movement, and therefore it was very easy from the beginning for us to achieve agreement between the parties.*
>
> *In this process, José Manuel Pureza, from the Bloc, was very important, because he was a bit of a pivot within Parliament between the various parties, and also the connection to the movement. And then, Isabel Moreira of the PS played a very important role, also in the coordination between the various proj- ects, and also with the movement.*
>
> *But there were different relationships. In other words, the relationship with the Greens was more distant. The relationship with the Liberal Initiative was practically non-existent. With PAN, it was also always close.*
>
> *Interviewed member of the civic movement 'Right to Die with Dignity'*

On the other hand, the representative of the civic movement against euthanasia mentioned that the fight in favour of euthanasia was mainly conducted through political parties and did not envisage the mobilization of civil society, as is evident in the following excerpt:

> *Because we saw all these movements also very much linked to the fulfilment of a certain world ideological agenda, etc. But when we look at the terrain here... On the eve of the debate of parliamentary votes on this law, we don't see anyone*

on the streets on the 'in favour' side. There are no demonstrations, no people are saying "I'm here to show my face". We only see politicians; we don't see anyone else. In other words, I've never seen mobilized physicians show their faces in favour of euthanasia, except for one or two, but they don't make much noise. I never saw lawyers either, nurses, or patients asking for euthanasia. What happens is that whenever these debates, these dates, take place, politicians and political parties will look for very specific, very individual cases and will mediatize them. So, they manage to put them in the greatest channel of visibility, which is television. (...) And they take on the volume and value almost of a crowd, when they are very specific cases.
    Interviewed member of the civic movement 'STOP Euthanasia'

8.3.2   Political Parties and Civil Society Movements Against Euthanasia

From another perspective, we questioned the interviewed members of civic movements if there was any kind of coordination between political parties and civic movements against the decriminalization of euthanasia.

In this sense, the representatives of the movement 'STOP Euthanasia' admitted having conversations with several MPs, to present their arguments. Moreover, they mentioned that their movement was completely non-partisan and without any kind of party or religious connection.

Without any kind of partisan or religious connection. (...) Because this is a theme that is relatable to all people, whether they are believers [in God] or not, whether they are from left or right [political parties]. I mean, this is a topic that is about being a person, of the human person. (...)
    I will just say one thing, which is the following: we learned in our visits to Parliament to look at others, not as someone who is on the other side or who is against it, but as human beings, who have a heart. And, therefore, our attitude is always heart-to-heart, and that is very beautiful to see. Because we feel, when we walk around, that we have no enemies there. We are very welcome when we ask to go to Parliament.
    Interviewed member of the civic movement 'STOP Euthanasia'

We also questioned the social actors with an opposing position if they felt some kind of coordenation between the political parties and the civic movements against the decriminalization of euthanasia. According to the representative of the movement 'Right to Die with Dignity', the political and social actors against decriminalization did not have a united front, as revealed in the following excerpt:

> *Obviously, there were some links from the CDS to the 'STOP Euthanasia' movement. But from the PCP there was none, and in fact, the feeling that we have had from this process so far was that, as with the Order of Physicians, the position of the majority of PCP militants does not coincide with the one that the leadership of the PCP took. (...) I think that, in terms of parties, they were not even close, nor organized, nor did they have the same arguments. I mean, some people in the PSD were together with the CDS and were part of the movement, eventually, but it was not even the majority of the PSD, I think.*
>
> *Interviewed member of the civic movement 'Right to Die with Dignity'*

NOTES

1. The Belém Palace is the official house of the President.
2. André Ventura is the leader of the Enough! party.

CHAPTER 9

Conclusions and Final Remarks

Abstract In this chapter, we summarize the main findings of this book and answer our main goals—to understand which political and social actors were most active in the political and public debate on euthanasia and which were the arguments and frames they mostly used to support their positions in favour and against the decriminalization of euthanasia.

The developed analysis enabled us to conclude that liberal political parties were the most important actors in defending the legalization of euthanasia, even if the civic movement 'Right to Die with Dignity' had played the important role of initiating the legislative process. The most often used arguments to support euthanasia were individual autonomy and compassion in ending insufferable pain.

As concerns the opposition to euthanasia, we concluded the most active actors were conservative political parties but also a multitude of organizational experts (particularly health-related ones) and interest groups, as well as the top echelon of the Portuguese Catholic Church. The most often used arguments were the danger of the 'slippery slope', the lack of palliative care and the inviolability of life.

These results led us to conclude euthanasia in Portugal can indeed be considered a morality issue, while this country can be considered part of the 'religious world' of morality politics.

Keywords Euthanasia; Portugal; Political parties; Conclusions

I. S. Almeida, L. F. Mota, *Politics and Policies in the Debate on Euthanasia*, https://doi.org/10.1007/978-3-031-44588-0_9

Chapters 1 and 2 explain that euthanasia may be considered a morality issue if its discussion involves the significant mobilization of morality-related arguments (Preidel & Knill, 2015). Considering the (apparent) simplicity of the issue and that it usually involves fundamental moral, ethical, philosophical and religious values, there is a tendency for a wider participation of actors in discussing it, including politicians but also several civil society actors (Knill, 2013). As described in Chap. 2, a recent trend of literature has identified the main factors that explain policy change and stability in morality issues, among which we may highlight political party cleavages, public opinion, the mobilization of civil society actors, and the role of courts and religion (Heichel et al., 2013; Knill, 2013). Recent literature (Engeli et al., 2012) has also demonstrated that there is a difference in trends of policy change in different groups of countries: On the one hand, party cleavages tend to be more important in countries from the so-called 'religious world' where confessional/conservative political parties have important places in Parliament and the religion-secular cleavage is therefore very evident; on the other hand, civil society actors tend to play a pivotal role in keeping morality issues in political agendas in countries from the 'secular world', as political parties do not have many political advantages in doing so.

After a long period without much debate, the issue of euthanasia has been increasingly under discussion in the so-called 'Western world', particularly since the 1980s (Preidel & Knill, 2015). This increasing debate about the issue is much related to the trends of secularization and individualization, but also to the emergence of different conceptions of the physician-patient relation or the increasing progress of medicine, which is nowadays capable of prolonging the lives of patients with severe illnesses (Preidel & Knill, 2015). This trend resulted in several countries permitting passive euthanasia and assisted suicide in recent decades, and, more recently, also permitting active euthanasia. Nevertheless, as of 2021, only seven countries had legalized active euthanasia, namely the Netherlands, Belgium (both since 2002), Luxembourg (since 2009), Colombia (since 2014), Canada (since 2016), Spain and New Zealand (both since 2021).

The political debate on active euthanasia officially started in Portugal in 2016 is an example of this trend. The legalization of active euthanasia in Portugal, even if consubstantiated only seven years after the beginning of this process, may nevertheless come as a surprise. Not only had Portugal legalized passive euthanasia quite recently (in 2012) but also has a comparatively high level of religiosity and societal conservatism, as results from the European Values Study demonstrate. However, this outcome is just one more step in the 'wave of legal permissiveness' on morality issues that the country has been dealing with since the early 2000s, with the decriminalization of drug use in 2001, the decriminalization of abortion in 2007, the legalization of same-sex marriage in 2010 and of the adoption by same-sex couples in 2016, the approval of self-determination of gender identity in 2018 or the approval of surrogacy in 2021.

The analysis of the debate on active euthanasia in Portugal was, therefore, an interesting object of study, not only for the reasons mentioned before but also because this debate was a particularly long and turbulent process, which lasted for seven years, comprised six rounds of discussion and involved several political and civil society actors using a multitude of arguments in favour of and against the legalization of euthanasia.

The goal of this book was then to analyse the political and public debate on euthanasia in Portugal so as to understand what the most active political and social actors in the debate were, either advocating the decriminalization of euthanasia or standing against it. Moreover, we aimed to understand if the arguments and frames used to support their positions were the same that had been used in other countries, and if they were mainly 'rationally-oriented' or 'morally-oriented'. More specifically, the aim of this book was to study the participation in the debate on the issue of euthanasia held in Portugal exercised by the following groups of actors: (i) political parties; (ii) veto players, such as the President and the Constitutional Court; (iii) experts; (iv) interest groups; (v) religious organizations, especially the Portuguese Catholic Church and (vi) other actors, namely those who decided to write opinion articles in newspapers.

One of the general conclusions was the confirmation that this legislative process was indeed very turbulent since it included two opposing petitions being widely discussed, one parliamentary rejection of bills by a small margin, two vetoes following declarations of unconstitutionality from the Constitutional Court related to unclear concepts and two political vetoes following decisions from the President, arguably also related to unclear parts of the proposed bills. Moreover, it also included the necessary coordination of bills from five different political parties and the strong opposition from four other political parties. Finally, this legislative process was particularly 'populated' by hearings from a huge number of actors—13 organizational experts, 21 individual experts[1] and 20 interest groups.

Regarding political parties, it could be concluded that those in favour of legalizing active euthanasia were libertarian political parties from the left wing (the Left Bloc, the Green Party, Free, most of the Socialist Party and the People-Animals-Nature party) and the right-wing (the Liberal Initiative). On the other hand, the parties against the legalization of euthanasia were conservative parties from the left wing (Communist Party) and the right-wing (most of the Social Democrat Party, the Popular Party and the Enough! party). It is also important to stress that two of these latter political parties—the Enough! party and, later, the Social Democrat Party—proposed the organization of a referendum, which was nevertheless rejected, but led to a more heated debate, as they were accused of trying to propose it merely as a political party strategy. Therefore, it may be concluded that the conservative-libertarian division was more important than the left-right division to understand the positions on the issue of euthanasia.

Moreover, we can also conclude that the religious/conservative vs. secular cleavage was important to understand this situation since, although only the Enough! party clearly evoked religious arguments, all the political parties against the legalization of euthanasia used arguments that fall into the category of conservative/traditional values. Likewise, the political parties which were in favour of euthanasia invoked the need to separate State practices from religious values, while accusing the political parties which voted against euthanasia of taking underlying religious values into consideration.

These conclusions are therefore in line with what was mentioned in Chaps. 1 and 2 about Portugal being a country from the 'religious world' (Engeli et al., 2012; Lago, 2023; Mota & Fernandes, 2022; Studlar et al., 2013, 2018). This assumption thus led us to hypothesize that political parties would have a stronger presence in the debate than civil society actors. This assumption was confirmed by the analysis in Chap. 8, in which we concluded that, although the legislative process had started with one petition from civil society actors, political parties in favour of euthanasia would nonetheless submit bills. Moreover, the interviewed actors who were against euthanasia consider that the process was dominated by political parties rather than civil society actors.

Although political parties were probably the most important actors for the legalization of active euthanasia, civil society actors were also very participative, not only submitting three petitions—one in favour of euthanasia, which started the legislative process, one against euthanasia and another which advocated the organization of a referendum—but also participating in the parliamentary debate through multiple hearings, written opinions requested by the parliamentary committee and other opinions issued on their own initiative. As mentioned before, 54 civil society actors participated in the debate—13 organizational experts, 21 individual experts and 20 interest groups.

Concerning the 13 organizational experts which participated in the debate, most followed request from the parliamentary committee which analysed the petitions and bills on euthanasia. The analysis in Chap. 5 revealed that most of the actors that fall into this category were against euthanasia, namely healthcare-related actors, such as the National Council of Ethics for Life Sciences, the Order of Physicians, the Order of Nurses, the director-general of health, the Portuguese Association of Palliative Care, the Portuguese Association of Bioethics and the Centre of Studies on Bioethics, as well as the Order of Lawyers. On the other hand, the remaining five organizational experts expressed a neutral position, namely the Order of Portuguese Psychologists, the High Council of the Judiciary, the High Council of the Public Prosecution Service, the Portuguese Association of Insurance Providers and the Specialty Council on Clinical and Health Psychology from the Order of Psychologists.

Regarding the 20 interest groups which participated in the parliamentary debate, three categories can be identified: three petitioners, nine interest groups dedicated to specific causes and eight religious interest groups. Moreover, it could be seen that eighteen of these interest groups expressed their opinion against the legalization of euthanasia—the Portuguese Federation for Life, which organized the petition against euthanasia; the subscribers of the Petition 48/XIV/1—Referendum on euthanasia; the civic movement 'STOP Euthanasia'; Portuguese Cáritas; the Association Together for Life; the Voiceless Children Movement; the civic movement 'Coimbra for Life'; the Hurray-for-Life Association; the Association InFamily; the Association for the Defence and Support of Life—Aveiro; the National Commission for Justice and Peace; the Association of Portuguese Catholic Physicians; the Interreligious Working Group on Religion & Health; the Portuguese Association of Catholic Jurists; the Portuguese Union of Seventh-day Adventists; the Portuguese Buddhist Union; the Portuguese Association of Catholic Psychologists and the Lisbon Jewish Community. On the other hand, the civic movement 'Right to Die with Dignity', which organized the first petition, was the only interest group in favour of legalizing euthanasia, while only the representative of the European Platform 'Wish to Die' expressed a neutral position.

The broad presence of interest groups in the discussion on euthanasia is therefore in line with the results from other studies about the agenda-setting and formulation of other morality issues, such as abortion and same-sex marriage, in which feminist and LGBTI+ organizations played a very important and active role (Alves et al., 2009; Mota & Fernandes, 2022; Santos, 2018a, b).

Our analysis also revealed that the Portuguese Catholic Church took an active stance regarding the discussion on euthanasia, with the submission of two pastoral letters and several notes and declarations. During the first round of discussion, besides these more traditional written notes, the Catholic Church was particularly active through the organization of the 'Week for Life', the distribution of leaflets and the signature of the Joint Declaration of Religious Confessions of the Interreligious Working Group on Religion & Heatlh—together with other Portuguese religious organizations such as the Evangelical Church, the Hindu community, the Islamic community, the Jewish community, the Buddhist community and the Ecumenical Patriarchate of Constantinople.

Therefore, this stance differs from the more passive one which the Catholic Church adopted regarding the discussions of other morality issues, such as abortion and same-sex marriage. This different posture is therefore in line with the decision of the current cardinal to adopt a more active posture regarding politics, particularly morality issues (Meyer-Resende & Hennig, 2015).

We also analysed the extra-parliamentary debate, namely the publication of opinion articles in one mainstream newspaper. As revealed in Chap. 7, 169 opinion articles (which covered the topic of euthanasia in a significant part of the article) were published between January 2016 and May 2023, most of them during the first and two rounds, particularly in crucial moments of the discussion—in March 2016, when the first petition was collecting signatures and was about to be submitted ($n=9$); in February 2017, when the second petition and the first bill were submitted ($n=7$); in May 2018, when the bills of the first round were voted on ($n=23$); in February 2020, when the five bills of the second round were discussed and approved ($n=37$); and, in March 2021, when the Constitutional Court declared the second-round bill to be unconstitutional and the President vetoed it ($n=10$). Moreover, it is important to highlight that the most active actors in publishing opinion articles were the scientific community

(23.67%), political actors (20.71%), health professionals (20.12%) and civil society actors (13.61%), and most of the opinion articles were against euthanasia (50.89%). In this regard, it is also important to stress that the actors who were mostly against euthanasia in their opinion articles were health professionals and political actors.

Finally, we would like to stress that there has been an evolution in public opinion on euthanasia. While the European Values Study results indicate that only 42.89% of the Portuguese population considered euthanasia justifiable, recent polls published in the media indicate that around two-thirds of the Portuguese population is now in favour of legalizing euthanasia.

As was concluded, political parties were the most relevant actors in favour of legalizing euthanasia, since all the other categories of involved actors (experts, interest groups, the Catholic Church and other civil society actors) were mostly against it. As for the arguments used, those most often heard against euthanasia were the danger of the slippery slope, the lack of palliative care and the inviolability of life, while the most often used to support euthanasia was individual autonomy and compassion in ending insufferable pain. This analysis thus led us to conclude that the arguments used in the discussion on euthanasia in Portugal are in line with the arguments mentioned in the literature (Ball, 2017; Keown, 2018; Lindsay, 2019; Preidel & Knill, 2015). Moreover, we may conclude that both moral and rational arguments are used by both 'sides' of the debate (Burlone & Richmond, 2018). Euthanasia in Portugal can therefore definitely be considered a morality issue.

NOTE

1. We decided not to analyze the hearings and written opinions from individual experts.

REFERENCES

Alves, M., Santos, A., Barradas, C., & Duarte, M. (2009). A despenalização do aborto em Portugal—Discursos, dinâmicas e acção coletiva: os referendos de 1998 e 2007 (The decriminalization of abortion in Portugal—Discourses, dynamics and collective action: The 1998 and 2007 referendums). *Oficina do CES no. 320.*

Ball, H. (2017). *The right to die: A reference handbook.* ABC-CLIO.

Burlone, N., & Richmond, R. (2018). Between morality and rationality: Framing end-of-life care policy through narratives. *Policy Sciences, 51*, 313–334.

Engeli, I., Green-Pedersen, C., & Larsen, L. (2012). The two worlds of morality politics—What have we learned? In E. I. Engeli, C. Green-Pedersen, L. T. Larsen, I. Engeli, C. Green-Pedersen, & L. T. Larsen (Eds.), *Morality politics in Western Europe* (pp. 185–199). Palgrave Macmillan.

Heichel, S., Knill, C., & Schmitt, S. (2013). Public policy meets morality: Conceptual and theoretical challenges in the analysis of morality policy change. *Journal of European Public Policy, 20*(3), 318–334. https://doi.org/10.108 0/13501763.2013.761497

Keown, J. (2018). *Euthanasia, ethics and public policy—An arguments against legalisation* (2nd ed.). Cambridge University Press.

Knill, C. (2013). The study of morality policy: Analytical implications from a public policy perspective. *Journal of European Public Policy, 20*(3), 309–317. https://doi.org/10.1080/13501763.2013.761494

Lago, I. (2023). Voting behaviour. In E. J. Fernandes, P. Magalhães, & A. Costa Pinto (Eds.), *The Oxford handbook of Portuguese politics* (pp. 276–290). Oxford University Press.

Lindsay, R. (2019). Euthanasia. In E. H. LaFollette (Ed.), *International encyclopedia of ethics*. John Wiley & Sons Ltd.

Meyer-Resende, M., & Hennig, A. (2015). Shunning direct intervention: Explaining the exceptional behaviour of the Portuguese Church hierarchy in morality politics. *New Diversities, 17*(1), 145–160.

Mota, L., & Fernandes, B. (2022). Debating the law of self-determination of gender identity in Portugal: Composition and dynamics of advocacy coalitions of political and civil society actors in the discussion of morality issues. *Social Politics: International Studies in Gender, State & Society, 29*(1), 50–70.

Preidel, C., & Knill, C. (2015). Euthanasia: Different moves towards punitive permissiveness. In E. C. Knill, C. Adam, & S. Hurka (Eds.), *On the road to permissiveness?: Change and convergence of moral regulation in Europe* (pp. 79–101). Oxford University Press.

Santos, A. (2018a). *A Institucionalização da Bioética e as Políticas Públicas de Saúde em Portugal (The institutionalization of bioethics and public health policies in Portugal)*. PhD thesis, Instituto Universitário de Lisboa, Lisboa.

Santos, A. C. (2018b). Luta LGBTQ em Portugal: duas décadas de histórias, memórias e resistências [LGBTQ struggle in Portugal: Two decades of stories, memories and resistance]. *Transversos: Revista de História, 14*, 37–52.

Studlar, D., Burns, G., & Cagossi, A. (2018). Morality policy processes in advanced industrial democracies. *Policy Studies, 39*(5), 479–497.

Studlar, D., Cagossi, A., & Duval, R. (2013). Is morality policy different? Institutional explanations for post-war Western Europe. *Journal of European Public Policy, 20*(3), 353–371.

REFERENCES

Álvarez, G., Kotera, Y., & Pina, J. (2022). *World Index of Moral Freedom—WIMF 2022*. Foundation for the Advancement of Liberty.

Accornero, G., & Pinto, P. (2023). Movements at the border. Conflict and protest in Portugal. In E. J. Fernandes, P. Magalhaes, & A. Pinto (Eds.), *The Oxford handbook of Portuguese politics* (pp. 457–471). Oxford University Press.

Adam, C., Knill, C., & Budde, E. (2020). How morality politics determine morality policy output—Partisan effects on morality policy change. *Journal of European Public Policy, 27*(7), 1015–1033.

Alves, M., Santos, A., Barradas, C., & Duarte, M. (2009). A despenalização do aborto em Portugal—Discursos, dinâmicas e acção coletiva: os referendos de 1998 e 2007 (The decriminalization of abortion in Portugal—Discourses, dynamics and collective action: The 1998 and 2007 referendums). *Oficina do CES no. 320.*

Ball, H. (2017). *The right to die: A reference handbook.* ABC-CLIO.

Birkland, T. (2016). *An introduction to the policy process theories, concepts, and models of public policy making* (4th ed.). Routledge.

Budde, E., Knill, C., Fernández-i-Marín, X., & Preidel, C. (2017). A matter of timing: The religious factor and morality policies. *Governance, 31*(1), 45–63.

Burlone, N., & Richmond, R. (2018). Between morality and rationality: Framing end-of-life care policy through narratives. *Policy Sciences, 51*, 313–334.

Cancela, J. (2023). Electoral Turnout. In E. J. Fernandes, P. Magalhaes, & A. Pinto (Eds.), *The Oxford handbook of Portuguese politics* (pp. 291–307). Oxford University Press.

I. S. Almeida, L. F. Mota, *Politics and Policies in the Debate on Euthanasia*, https://doi.org/10.1007/978-3-031-44588-0

Cohen, J., Van Landeghem, P., Carpentier, N., & Deliens, L. (2014). Public acceptance of euthanasia in Europe: A survey study in 47 countries. *International Journal of Public Health, 59*, 143–156.

Cohen, M., March, J., & Olsen, J. (1972). A Garbage can framework of organizational choice. *Administrative Science Quarterly, 17*(1), 1–25.

Davies, H., Nutley, S., & Smith, P. (2000). *What works? Evidence-based policy and practice in public services.*

Emanuel, E., Onwuteaka-Philipsen, B., Urwin, J., & Cohen, J. (2016). Attitudes and practices of euthanasia and physician-assisted suicide in the United States, Canada, and Europe. *Clinical Review & Education, 36*(1), 79–90.

Engeli, I., Green-Pedersen, C., & Larsen, L. (2012a). Introduction. In E. I. Engeli, C. Green-Pedersen, & L. T. Larsen (Eds.), *Morality politics in Western Europe: Parties, agendas and policy choices* (pp. 1–4). Palgrave Macmillan.

Engeli, I., Green-Pedersen, C., & Larsen, L. (2012b). The two worlds of morality politics—What have we learned? In E. I. Engeli, C. Green-Pedersen, L. T. Larsen, I. Engeli, C. Green-Pedersen, & L. T. Larsen (Eds.), *Morality politics in Western Europe* (pp. 185–199). Palgrave Macmillan.

Escada, M., & Lucas, T. (2019). A desregulação profissional em Portugal durante a Troika: o caso da Ordem dos Médicos e dos Advogados (Professional deregulation in Portugal during the Troika: The case of Physicians and Lawyers Professional Associations). In E. M. Lisi & M. Lisi (Eds.), *Grupos de Interesse e Crise Económica em Portugal (Interest groups and economic crisis in Portugal)* (pp. 207–236). Edições Sílabo.

Euchner, E.-M. (2019). *Morality politics in a secular age: Strategic parties and divided governments in Europe.* Palgrave Macmillan.

Félix, Z., Costa, S., Alves, A., Andrade, C., Duarte, M., & Brito, F. (2013). Eutanásia, distanásia e ortotanásia: uma revisão integrativa da literatura (Euthanasia, dysthanasia and orthothanasia: An integrative literature review). *Ciência & Saúde Coletiva, 18*(9), 2733–2746. https://doi.org/10.1590/S1413-81232013000900029

Fontalis, A., Prousali, E., & Kulkarni, K. (2018). Euthanasia and assisted dying: What is the current position and what are the key arguments informing the debate? *Journal of the Royal Society of Medicine, 111*(11), 407–413.

Franco, R. (2015). *Diagnóstico das ONG em Portugal.* Fundação calouste Gulbenkian.

Freire, A. (2023). The centre-left and the radical left in Portuguese democracy, 1974–2021. In E. J. Fernandes, P. Magalhaes, & A. Pinto (Eds.), *The Oxford handbook of Portuguese politics* (pp. 88–101). Oxford University Press.

Garoupa, N., & Tiede, L. (2023). Judicial politics in Portugal. In E. J. Fernandes, P. Magalhaes, & A. Pinto (Eds.), *The Oxford handbook of Portuguese politics* (pp. 164–180). Oxford University Press.

Goes, E., & Leston-Bandeira, C. (2023). The role of the Portuguese parliament. In E. J. Fernandes, P. Magalhaes, & A. Pinto (Eds.), *The Oxford handbook of Portuguese politics* (pp. 136–148). Oxford University Press.

Haider-Markel, D., & Meier, K. (1996). The politics of gay and lesbian rights: Expanding the scope of the conflict. *Journal of Politics, 58*(2), 332–349.

Heclo, H. (1978). Issue networks and the executive establishment. In E. A. King (Ed.), *The new American political system* (pp. 87–124). American Enterprise Institute Press.

Heichel, S., Knill, C., & Schmitt, S. (2013). Public policy meets morality: Conceptual and theoretical challenges in the analysis of morality policy change. *Journal of European Public Policy, 20*(3), 318–334. https://doi.org/10.108 0/13501763.2013.761497

Howlett, M., Ramesh, M., & Perl, A. (2020). *Studying public policy: Principles and processes* (4th ed.). Oxford University Press. Obtido de https://global.oup.com/academic/product/studying-public-policy-9780199026142?q=Michael%20 Howlett&lang=en&cc=pt#

Hurka, S., Adam, C., & Knill, C. (2017). Is morality policy different? Testing sectoral and institutional explanations of policy change. *Policy Studies Journal, 45*(4), 688–712.

Inbadas, H., Zaman, S., Whitelaw, S., & Clark, D. (2017). Declarations on euthanasia and assisted dying. *Death Studies, 41*(9), 574–584.

Jalali, C., & Teruel, J. (2019). Parliamentary party groups in the Iberian democracies. In E. J. Fernandes & C. Leston-Bandeira (Eds.), *The Iberian legislatures in comparative perspective* (pp. 49–70). Routledge.

Jann, W., & Wegrich, K. (2007). Theories of the policy process. In E. F. Fischer, G. Miller, & M. Sidney (Eds.), *Handbook of public policy analysis: Theory, politics, and methods* (pp. 43–62). CRC Press.

Jordan, A. (1981). Iron triangles, Woolly corporatism and elastic nets: Images of the policy process. *Journal of Public Policy, 1*(1), 95–124.

Kane, L. (2014). *Medscape ethics report 2014, Part 1: Life, death, and pain*. Obtido de Medscape website: http://www.medscape.com/features/slideshow/public/ethics2014-part1

Keown, J. (2018). *Euthanasia, ethics and public policy   An arguments against legalisation* (2nd ed.). Cambridge University Press.

Knill, C. (2013). The study of morality policy: Analytical implications from a public policy perspective. *Journal of European Public Policy, 20*(3), 309–317. https://doi.org/10.1080/13501763.2013.761494

Knill, C., Adam, C., & Hurka, S. (2015). *On the road to permissiveness? Change and convergence of moral regulation in Europe*. Oxford University Press.

Knill, C., & Tosun, J. (2020). *Public policy: A new introduction* (2nd ed.). Palgrave Macmillan. Obtido de https://www.macmillanihe.com/page/detail/public-policy-christoph-knill/?k=9780230278387

Kouwenhoven, P., Raijmakers, N., van Delden, J., Rietjens, J., Schermer, M., van Thiel, G., et al. (2012). Opinions of health care professionals and the public after eight years of euthanasia legislation in the Netherlands: A mixed methods approach. *Palliative Medicine, 27*(3), 273–280.

Lago, I. (2023). Voting behaviour. In E. J. Fernandes, P. Magalhães, & A. Costa Pinto (Eds.), *The Oxford handbook of Portuguese politics* (pp. 276–290). Oxford University Press.

Lindblom, C. (1959). The science of muddling through. *Public Administration Review, 19*(2), 79–88.

Lindsay, R. (2019). Euthanasia. In E. H. LaFollette (Ed.), *International encyclopedia of ethics.* John Wiley & Sons Ltd.

Lisi, M., & Loureiro, J. (2023). Interest groups, business associations and unions. In E. J. Fernandes, P. Magalhaes, & A. Pinto (Eds.), *The Oxford handbook of Portuguese politics* (pp. 423–439). Oxford University Press.

Lisi, M., & Marquez, L. (2019). Interest groups in the Iberian parliaments. In E. J. Fernandes & C. Leston-Bandeira (Eds.), *Iberian legislatures in comparative perspective* (pp. 130–148). Routledge.

Magalhaes, P. (2023). Citizens and politics: Support and engagement. In E. J. Fernandes, P. Magalhaes, & A. Pinto (Eds.), *The Oxford handbook of Portuguese politics* (pp. 244–261). Oxford University Press.

Marchi, R., & Alves, A. (2023). The right and far-right in the Portuguese democracy (1974–2022). In E. J. Fernandes, P. Magalhaes, & A. Pinto (Eds.), *The Oxford handbook of Portuguese politics* (pp. 102–118). Oxford University Press.

Meyer Resende, M. (2023). The relations between the Catholic church and the political arena in Portugal. In E. J. Fernandes, P. Magalhaes, & A. Pinto (Eds.), *The Oxford handbook of Portuguese politics* (pp. 472–486). Oxford University Press.

Meyer-Resende, M., & Hennig, A. (2015). Shunning direct intervention: Explaining the exceptional behaviour of the Portuguese Church hierarchy in morality politics. *New Diversities, 17*(1), 145–160.

Moniz, J. (2018). Índice de Religiosidade: Uma proposta de teorização e medição dos fenómenos religiosos contemporâneos (Religiosity Index: A proposal for theorizing and measuring contemporary religious phenomena). *Revista Brasileira de História das Religiões, 32*, 191–219.

Mota, L., & Fernandes, B. (2022). Debating the law of self-determination of gender identity in Portugal: Composition and dynamics of advocacy coalitions of political and civil society actors in the discussion of morality issues. *Social Politics: International Studies in Gender, State & Society, 29*(1), 50–70.

Nebel, K., & Hurka, S. (2015). Abortion: Finding the impossible compromise. In E. C. Knill, C. Adam, & S. Hurka (Eds.), *On the road to permissiveness: Change and convergence of moral regulation in Europe* (pp. 58–78). Oxford University Press.

Neto, O. (2023). Semi-presidentialism in Portugal: Academic quarrels amidst institutional stability. In E. J. Fernandes, P. Magalhaes, & A. Pinto (Eds.), *The Oxford handbook of Portuguese politics* (pp. 121–135). Oxford University Press.

Peters, B., & Pierre, J. (2006). Introduction. In E. B. G. Peters & J. Pierre (Eds.), *Handbook of public policy* (pp. 1–9). Sage.

Pinto, A., & Paris, A. (2023). Democratization and its legacies. In E. J. Fernandes, P. Magalhaes, & A. Pinto (Eds.), *The Oxford handbook of Portuguese politics* (pp. 18–37). Oxford University Press.

Pratas, M., & Bizzarro, F. (2023). Political parties and party system. In E. J. Fernandes, P. Magalhaes, & A. Pinto (Eds.), *The Oxford handbook of Portuguese politics* (pp. 353–370). Oxford University Press.

Preidel, C., & Knill, C. (2015). Euthanasia: Different moves towards punitive permissiveness. In E. C. Knill, C. Adam, & S. Hurka (Eds.), *On the road to permissiveness?: Change and convergence of moral regulation in Europe* (pp. 79–101). Oxford University Press.

Sabatier, P. (1988). An advocacy coalition framework of policy change and the role of policy-oriented learning therein. *Policy Sciences, 21*, 129–168.

Sabatier, P., & Weible, C. (2007). The advocacy coalition framework innovations and clarifications. In E. P. Sabatier (Ed.), *Theories of the policy process* (2nd ed., pp. 189–220). Routledge.

Salgado, S. (2023). Mass media and political communication. In E. J. Fernandes, P. Magalhaes, & A. Pinto (Eds.), *The Oxford handbook of Portuguese politics* (pp. 308–321). Oxford University Press.

Santos, A. (2018). *A Institucionalização da Bioética e as Políticas Públicas de Saúde em Portugal (The institutionalization of bioethics and public health policies in Portugal)*. PhD thesis, Instituto Universitário de Lisboa, Lisboa.

Santos, A. C. (2018). Luta LGBTQ em Portugal: duas décadas de histórias, memórias e resistências [LGBTQ struggle in Portugal: Two decades of stories, memories and resistance]. *Transversos: Revista de História, 14*, 37–52.

Schmitt, S., Euchner, E.-M., & Preidel, C. (2013). Regulating prostitution and same-sex marriage in Italy and Spain: The interplay of political and societal veto players in two catholic societies. *Journal of European Public Policy, 20*(3), 425–441.

Silva, P. (2020). *Jobs for the Boys? Nomeações para a administração pública*. Fundação Francisco Manuel dos Santos.

Silva, S., Azevedo, L., & Ricou, M. (2019). Determinantes na opinião sobre eutanásia em amostra de médicos portugueses. *Revista Iberoamericana de Bioética, 10*, 1–19.

Stone, D. (2007). Public policy analysis and think tanks. In E. F. Fischer, G. Miller, & M. Sidney (Eds.), *Handbook of public policy analysis: Theory, politics, and methods* (pp. 149–157). CRC Press.

Studlar, D., Burns, G., & Cagossi, A. (2018). Morality policy processes in advanced industrial democracies. *Policy Studies, 39*(5), 479–497.

Studlar, D., Cagossi, A., & Duval, R. (2013). Is morality policy different? Institutional explanations for post-war Western Europe. *Journal of European Public Policy, 20*(3), 353–371.

Tavares, A. (2019). *Administração pública portuguesa*. Fundação Francisco Manuel dos Santos.

Weible, C., & Nohrstedt, D. (2012). The advocacy coalition framework: Coalitions, learning and policy change. In E. E. Araral Jr., S. Fritzen, M. Howlett, M. Ramesh, & X. Wu (Eds.), *Routledge handbook of public policy* (pp. 125–137). Routledge.